W9-DAQ-390

Handbook of
Sound System Design

by John Eargle

ELAR Publishing Co. ,Inc.
Commack, NY 11725

First Edition

ELAR Publishing Company Inc.
Commack, New York 11725
Copyright© 1989

All rights reserved. No part of this
publication, or parts thereof, may be
reproduced in any form without the
permission of the copyright owner.

Printed in the United States of America
Library of Congress Catalog Number 88-82371
ISBN Number 0-914130-03-X

First Printing

Illustrations by Hal Keith
Page layout by K&S Graphics, Island Park, N.Y.

Contents

PREFACE

Over recent years, sound system design has reached a high level of sophistication as system designers and component manufacturers have worked to solve their common problems. Education is an important part in this process, and the manufacturers of components used in professional sound system design routinely present product usage seminars aimed at informing the designer of new products and their application. Feedback from demanding users has also played an important role in advanced product development, and there are few manufacturers today who do not rely on such input.

Independent educational activities continue to play an important role in introducing young practitioners to the field, and an increasing number of schools now teach the rudiments of sound system design. The Handbook of Sound System Design is aimed at both the seasoned designer and the beginner in the field, and as such can be used as an educational text. Its use as a handbook in the traditional sense is facilitated through extensive indexing.

The book grew out of a series of articles on sound reinforcement which appeared in **db Magazine**. At the time, editor Larry Zide expressed interest in developing the series into a handbook. As our thoughts came together, it was decided to expand the coverage to include basic tutorial chapters covering electrical fundamentals, acoustical fundamentals, and psychoacoustical fundamentals, as well as coverage of motion-picture sound and high-level music reinforcement. The primary thrust of the book, however, remains sound reinforcement.

The book is organized for self study. The first three chapters provide necessary background in electrical and acoustical fundamentals, and the coverage then moves on to an intensive study of the basic hardware elements used in sound system design: low, mid, and high-frequency devices, dividing networks, microphones, and the many electronic devices used in system architecture and layout.

The book then turns to detailed treatment of acoustical gain, sound system equalization practice, and intelligibility estimation.

The final nine chapters deal with specific areas of application, as given below: central loudspeaker array design, distributed array design, paging systems, artificial ambience systems, speech privacy and noise masking systems, high-level sound reproduction, sound in the theater, music reinforcement, and line arrays.

The author is grateful to the many manufacturers and workers in the field who have provided information used in the book. Credits are given as this information is presented. Special thanks are due to JBL Professional for permission to use many illustrations from their Sound System Design Reference Manual.

The author wishes to acknowledge John Hoge, who studied the outline and manuscript of the book and made many helpful suggestions.

John Eargle
September 1988

Electrical Fundamentals

INTRODUCTION

A knowledge of electrical fundamentals is essential for the practice and application of sound reinforcement at any level. The subject is a vast one, and our coverage here will be limited to topics of prime concern in sound reinforcement, namely, the interaction between electronic systems and electrical loads.

In this chapter we will cover power relationships, impedance, resonance, decibel notation, transformation, attenuators, and wire losses. Later chapters will elaborate, as need be, on other electrical topics.

ELECTRICAL POWER; OHM'S LAW

In direct current (DC) circuits, the following quantities and units are encountered:

Quantity:	Unit:	Symbol:
Power	Watt	P
Current	Ampere	I
Potential (Voltage)	Volt	E or V
Resistance	Ohm (Ω)	R

Figure 1-1 shows a simple series circuit containing a battery E and a load resistance R. If we place an ammeter (a device for measuring current) in series with the load, and place a voltmeter (a device for measuring potential) in parallel with the load, we can calculate the power delivered to the load as:

$$P = E \times I \tag{1-1}$$

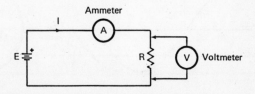

Figure 1-1. Direct current-power relationships.

The potential drop across the load will be equal to the battery potential, and that drop can be expressed in a form known as Ohm's Law:

$$E = I \times R \tag{1-2}$$

These two expressions can be combined in a number of ways:

$$P = EI = I^2R = E^2/R \tag{1-3}$$

$$E = IR = P/I = \sqrt{PR} \tag{1-4}$$

$$I = E/R = P/E = \sqrt{P/R} \tag{1-5}$$

$$R = E/I = P/I^2 = E^2/P \tag{1-6}$$

Knowing any two of these quantities, we can easily calculate the other two.

The preceding example used a battery as a dc power source. Alternating current, or ac, sources are more commonly encountered today in power transmission. All audio signals are made up of alternating current components.

The sine wave is the fundamental component of alternating current, and it is shown in Figure 1-2A. The sine wave is a periodic function; that is, it repeats itself at time intervals, or cycles, known as the period of the waveform. Lower-case t is the symbol for *period*, and it is related to *frequency* as:

Frequency (Hz) = 1/t, where t is time in seconds. (1-7)

The average value of a sine wave is 0.637 times its peak value, and

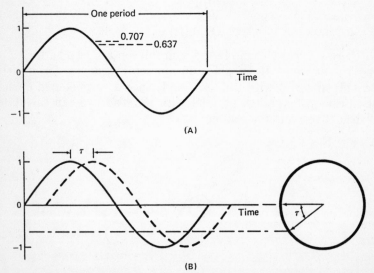

Figure 1-2. The sine wave.

its RMS (Root Mean Square) value is $1/\sqrt{2}$, or 0.707 times its peak value. The RMS value represents the effective value of the current waveform in power calculations. That is, an alternating current of one ampere peak value will produce the same power in a load resistor as would a direct current of 0.707 amperes. Ammeters and voltmeters are normally calibrated to read RMS values of sine waves unless otherwise indicated. The circuit of Figure 1-1 and the equations developed from Ohm's Law hold for ac as well as dc signals.

Figure 1-2B shows another important aspect of periodic waveforms, *phase* relationships. Here, we see two sine waves of the same frequency. Note that the dotted one lags the solid one by a time interval, τ. The phase relationship is often stated in degrees, with 360 degrees representing one period of the waveform.

IMPEDANCE:

The most common electrical load encountered in sound reinforcement is the loudspeaker. While the loudspeaker is largely a resistive load, it contains mechanical elements (mass and compliance) which behave, electrically, at the loudspeaker's terminals as inductive and capacitive elements. While these elements inhibit the flow of current, they are not resistive; rather, they are *reactive*, storing energy rather than dissipating it. Thus, an inductor exhibits *inductive reactance*, and a capacitor exhibits *capacitive reactance*.

The combined effect of resistance and the two kinds of reactance is called *impedance*. For the most part, we can treat impedance just the way we did resistance earlier in this chapter. The symbol Z represents impedance, and we can simply replace the R in equations 3 through 6 with Z.

The symbol for reactance is X. Like resistance, it is measured in ohms. But there is an important difference; resistance is independent of frequency, whereas reactance is a function of frequency. Inductive reactance is given by:

$$X_L = j\omega L, \tag{1-8}$$

where L is the inductance measured in henrys, ω is equal to $2\pi f$, and j is the so-called complex operator (the square root of –1), signifying a phase rotation of 90 degrees.

The significance of the j-operator may be explained as follows. If a sine wave potential is applied across a resistor, a sine wave current will flow through the resistor which is *in phase* with the applied potential. If the sine wave potential is applied across an inductor, the resulting current will *lag* the applied potential by 90 degrees. Furthermore, with each doubling of frequency, the reactance of the inductor, measured in ohms, will double.

Capacitive reactance is given by:

$X_C = 1/(j\omega C) = -j/(\omega C)$.

where C is the capacitance measured in farads.

In the capacitive case, the negative j-operator indicates that current through the capacitor will *lead* the applied voltage by 90 degrees. Furthermore, the reactance of the capacitor will drop by one-half for each doubling of frequency.

Note that X_L is proportional to frequency and X_C is inversely proportional to frequency.

COMBINING LOADS:

Electrical loads may be combined in series or in parallel, as shown in Figure 1-3A and B. When the loads are all of equal impedance, it is easy to calculate the resulting impedance, as shown in the figures. However, when the impedances are different, then the following equations must be used:

For loads in series:

$$Z_T = Z_1 + Z_2 + Z_3 + \ldots + Z_n \tag{1-9}$$

For loads in parallel:

$$Z_T = \frac{1}{1/Z_1 + 1/Z_2 + 1/Z_3 + \ldots + 1/Z_n} \tag{1-10}$$

For the special case of only two different values of impedance, the parallel combination is given by the following equation:

$$Z_T = \frac{Z_1 Z_2}{Z_1 + Z_2} \tag{1-11}$$

RESONANCE:

In the circuit shown in Figure 1-4A, a variable frequency source, E, is placed in series with a capacitor, inductor, and resistor. At very low frequencies, the value of X_C will be extremely high, and the current flowing through the circuit will be very low. Thus, the potential drop

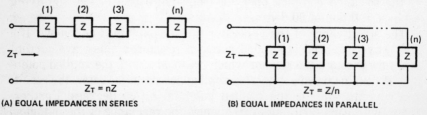

(A) EQUAL IMPEDANCES IN SERIES   (B) EQUAL IMPEDANCES IN PARALLEL

Figure 1-3. Electrical loads in series and parallel.

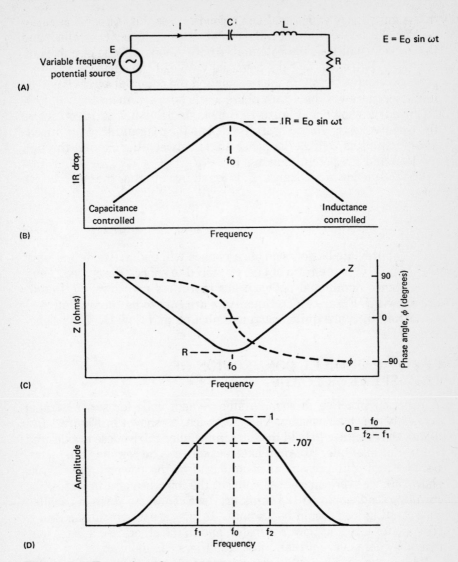

Figure 1-4. Examples of resonance.

across the load resistor will be small. At very high frequencies, the value of X_L will be very high, resulting in a low value of current and potential drop across the load resistor.

The impedance of this circuit is:

$$Z_T = R + j(\omega L - 1/\omega C)$$

Now, at some frequency between these extremes, the values of ωL and $1/\omega C$ will be equal. They will cancel, leaving R as the only term

5

in the impedance equation. Such a condition as this is known as series resonance. Below resonance, the circuit is capacitance controlled, and the current will lead the potential. Above resonance, the circuit is inductance controlled, and the potential will lead the current. At resonance, the current through the resistor and the potential across it will be in phase; there will be a zero phase angle between them.

The curve shown at Figure 1-4B plots the IR (potential) drop across the resistive load, while Figure 1-4C plots the magnitude of the impedance in ohms as well as the phase angle between the current through the load and the potential across it.

The frequency of resonance, f_o, is found by equating the two reactive terms:

$$f_o = \frac{1}{2\pi\sqrt{LC}} \tag{1-12}$$

Our abbreviated discussion of resonance will end with a definition of the term Q. Q is the ratio of energy stored to energy dissipated. Thus, a resistance dominated network has more loss and lower Q. Details are shown in Figure 1-4D. Equalizers are said to be "high-Q" if their peaks and dips are quite sharp and abrupt, and "low-Q" if the slopes are gentle.

SERIES-PARALLEL CONNECTION OF LOUDSPEAKER LOADS

A common practice in sound reinforcement calls for series-parallel connection of loudspeakers. This approach, as shown in Figure 1-5A, allows the designer to hold the total impedance fairly close to a desired value. For example, 16 8-ohm loudspeakers can be combined in series-parallel and still present an 8-ohm load to the power amplifier, as shown. If we were merely to connect the loudspeakers in series, the resulting load would be 128 ohms; or, if we connected them in parallel, the resulting load would be 0.5 ohms. In one case, the resulting load is too high for efficient power transfer, and in the other case it is too low, drawing more current than most amplifiers can deliver.

Where the number of loudspeakers of the same impedance to be combined is equal to a perfect square (4, 9, 16, 25, etc.), then series-parallel loading can always result in a load whose impedance equals that of a single loudspeaker. Where the number to be combined is not a perfect square, then some compromise must be made. For example, in Figure 1-5B, we show 12 loudspeakers grouped as four in series, three in parallel, with a resulting impedance of 6 ohms. In 1-5C, we show three in series, four in parallel, with a resulting impedance of 10⅔ ohms. As we will see in later chapters, the choice here has to do with the characteristics of power amplifiers and their output ratings for reduced load impedances.

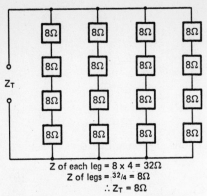

Z of each leg = 8 x 4 = 32Ω
Z of legs = $^{32}/_4$ = 8Ω
∴ Z_T = 8Ω

(A) EQUAL IMPEDANCES IN SERIES-PARALLEL,
16 LOUDSPEAKERS

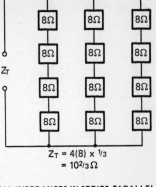

Z_T = 4(8) x ⅓
= 10⅔Ω

(B) EQUAL IMPEDANCES IN SERIES-PARALLEL,
12 LOUDSPEAKERS

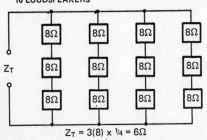

Z_T = 3(8) x ¼ = 6Ω

(C) EQUAL IMPEDANCES IN SERIES-PARALLEL,
12 LOUDSPEAKERS

Figure 1-5. Series-parallel connection of loudspeakers.

SERIES AND PARALLEL COMBINING OF RESISTORS, INDUCTORS, AND CAPACITORS

Equations similar to the ones we presented for combining impedances can be used for combining values of resistance or inductance:

For n resistors in series:

$$R_T = R_1 + R_2 + R_3 + \ldots + R_n \tag{1-13}$$

For n resistors in parallel:

$$R_T = \frac{1}{1/R_1 + 1/R_2 + 1/R_3 + \ldots + 1/R_n} \tag{1-14}$$

For two resistors in parallel:

$$R_T = \frac{R_1 R_2}{R_1 + R_2} \tag{1-15}$$

For n inductors in series:

$$L_T = L_1 + L_2 + L_3 + \ldots + L_n \tag{1-16}$$

7

For n inductors in parallel:

$$L_T = \frac{1}{1/L_1 + 1/L_2 + 1/L_3 + \ldots + 1/L_n} \tag{1-17}$$

For two inductors in parallel:

$$L_T = \frac{(L_1)(L_2)}{L_1 + L_2} \tag{1-18}$$

For capacitance, we invert the equations. For n capacitors in series:

$$C_T = \frac{1}{1/C_1 + 1/C_2 + 1/C_3 + \ldots + 1/C_n} \tag{1-19}$$

For n capacitors in parallel:

$$C_T = C_1 + C_2 + C_3 + \ldots + C_n \tag{1-20}$$

For two capacitors in series:

$$C_T = \frac{C_1 C_2}{C_1 + C_2} \tag{1-21}$$

THE DECIBEL:

In sound reinforcement, we deal with an extremely wide range of powers and voltages. These quantities are often unwieldy, and decibel notation is a convenient way to reduce large numbers to smaller ones. The term *level* is usually used for ratios expressed in decibels.

The *bel* is defined as the common logarithm of a power ratio:

Level difference in *bels* = $\log (P_1/P_0)$, where P_0 is some reference power.

The *decibel* (dB) is one-tenth *bel*; thus:

Level difference in dB = $10 \log (P_1/P_0)$ (1-22)

Letting P_0 be one watt, we can construct the following table:

P_1 (watts)	Level in dB
0.01	-20
0.1	-10
1	0
10	10
100	20

The reader can see the pattern here; each ten-to-one ratio in power represents a level change of 10 dB. Note that the entire power range of 10,000-to-1 is represented by a level difference of 40 dB.

Constructing another table:

P_1 (watts)	Level in dB
0.25	–6
0.5	–3
1	0
2	3
4	6

The pattern observed here is that each two-to-one ratio in power corresponds to a level difference of 3 dB.

In the following table, we will pick power ratios that result in one dB level steps:

P_1 (watts)	Level in dB
0.1	–10
0.125	–9
0.16	–8
0.2	–7
0.25	–6
0.315	–5
0.4	–4
0.5	–3
0.63	–2
0.8	–1
1	0
1.25	1
1.6	2
2	3
2.5	4
3.15	5
4	6
5	7
6.3	8
8	9
10	10

A few exercises will help the reader gain the familiarity he needs.

1. What power level, relative to one watt, is represented by 50 watts? From the preceding table, we see that 5 watts represent a level of 7 dB. Since 50 is 10 times 5, this represents another 10 dB. Therefore, $7 + 10 = 17$ dB.

2. What is the level difference between 25 and 80 watts? From the above table, we see that the level corresponding to 2.5 is 4 dB, and the level corresponding to 8 is 9 dB. The difference is $9 - 4 = 5$ dB. Since the ratio of 2.5 and 8 is the same as the ratio of 25 and 80, the same 5 dB is our answer.

3. An amplifier has an input impedance of 600 ohms and is terminated with a load of 8 ohms. An input of .775 volts results in an output of 40 volts. What is the power gain of the amplifier? We must calculate both input and output powers and take 10 log of their ratio. Recalling the equations we developed earlier:

Input Power $= .775^2/600\Omega = 0.001$ watt
Output Power $= 40^2/8\Omega = 200$ watts
Power Ratio $= 200/.001 = 200,000$

Noting that all 10-to-1 power ratios correspond to level differences of 10 dB, a 100,000-to-1 power ratio represents a level difference of 50 dB. The remaining ratio, 200,000-to-100,000, is simply a 2-to-1 ratio, and this corresponds to a level difference of 3 dB. Therefore, $50 + 3 = 53$ dB.

Figure 1-6 presents a convenient nomograph for determining level differences by inspection. Simply locate the two powers on the scale and read the level difference between them directly in dB.

A level difference in dB can be converted back to a power ratio with the following equation:

$$\text{Power Ratio} = 10^{\frac{dB}{10}} \qquad (1\text{-}23)$$

As an example, find the power ratio corresponding to a level difference of 16 dB.

$$10^{\frac{16}{10}} = 39.8$$

As a rule, we measure potential and current, not power, in electrical circuits, and it has become common to use potential ratios in determining level differences in dB. This can be done provided that the potentials are measured across the same impedance. The equation we use for this is:

$$\text{Level difference in dB} = 20 \log (E_1/E_0) \qquad (1\text{-}24)$$

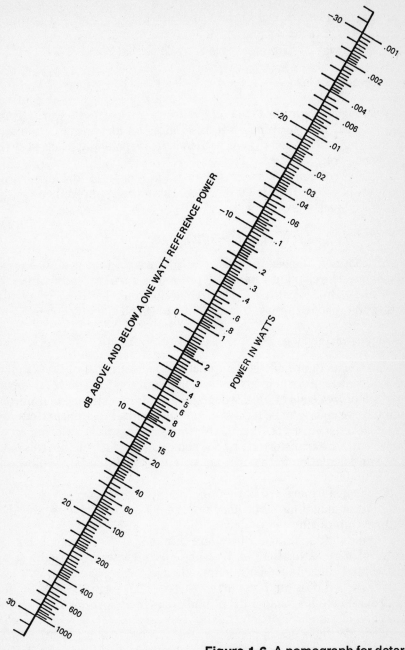

Figure 1-6. A nomograph for determining power ratios in decibels.

11

The reason for the factor of 20 will be apparent from Figure 1-7. At A, a power of one watt is dissipated in the load. When the source potential is doubled, the current doubles as well, and the power increases by a factor of four.

Another way of looking at this is by the equation:

$$\text{Level} = 10 \log \left(\frac{E_1^2}{Z} \times \frac{Z}{E_0^2} \right) = 10 \log \left(\frac{E_1}{E_0} \right)^2 = 20 \log \frac{E_1}{E_2} \quad (1\text{-}25)$$

A nomograph similar to that of Figure 1-6 is shown in Figure 1-8 and can be used to determine level differences by inspection of potential ratios. The designation dBV is used to indicate potential levels above or below one volt.

Another reference power used in some phases of low-level signal transfer is one milliwatt (.001 watt). The normal impedance for this reference level is 600 ohms, thus:

$$E = \sqrt{PZ} = \sqrt{(.001)(600)} = 0.775 \text{ volts}$$

It is common to measure levels in dBm by noting the potential across 600 ohms. The chart of Figure 1-9 presents a small range of values for dBV and dBm. Recalling the 20 dB level change for 10-to-1 potential ratios, the reader can use this table well beyond the given range.

TRANSFORMERS

Transformers are a useful adjunct to ac and audio circuitry. Details of an ideal transformer are shown in Figure 1-10. The transformer consists of two coils of wire wound on a common laminated iron core, and the two coils may be tapped at various points. Potential and current are transformed according to the turns ratio, while impedance relationships are transformed by the square of the turns ratio. Note that a power calculation in the primary winding yields the same value as in the secondary.

Real transformers are limited in a number of ways:

1. Power handling. Transformers are not lossless devices, and they will heat up during heavy drive.

2. Bandwidth. Transformers are limited at low frequencies because of core saturation, and at high frequencies because of stray, distributed capacitance between windings.

3. Insertion loss and distortion. Good transformers will exhibit insertion losses less than 1 dB at mid-frequencies. At low frequencies, insertion loss and distortion usually increase.

Outside of their use in power supplies, transformers are often found in the input stages of microphone preamplifiers and in line distribution systems for paging purposes.

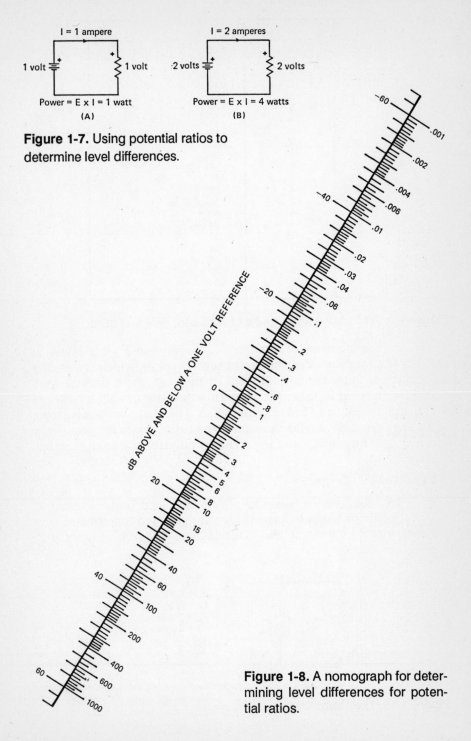

Figure 1-7. Using potential ratios to determine level differences.

Figure 1-8. A nomograph for determining level differences for potential ratios.

POTENTIAL	dB re 1 VOLT (dBV)	POTENTIAL	POWER LEVEL, dBm
3.15	10	2.45	10
2.80	9	2.20	9
2.50	8	1.95	8
2.23	7	1.73	7
2.00	6	1.55	6
1.80	5	1.38	5
1.60	4	1.23	4
1.40	3	1.10	3
1.25	2	.98	2
1.12	1	.87	1
1.00	0	.775	0
.90	−1	.69	−1
.80	−2	.62	−2
.70	−3	.55	−3
.63	−4	.49	−4
.56	−5	.44	−5
.50	−6	.39	−6
.45	−7	.35	−7
.40	−8	.31	−8
.35	−9	.27	−9
.315	−10	.245	−10

Figure 1-9. A chart of a small range of values for dBV and dBm.

The autotransformer, or autoformer, is shown in Figure 1-11A. It consists of a single winding tapped at several points, which are usually at convenient impedance multiples. In the past, these devices were used in engineered loudspeaker arrays to adjust the various drive levels and match impedances to the amplifier. A typical application is shown at B. They are used less for these purposes today, due to the prevalence of solid state amplifiers and the rising popularity of biamplification.

ATTENUATORS (pads)

Pads are often used between amplifiers and loudspeakers to adjust the relative drive levels to various high-frequency components. There are two kinds in general use, "L-pads" and "T-pads." Figure 1-12

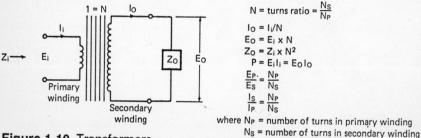

N = turns ratio = $\dfrac{N_S}{N_P}$

$I_O = I_i / N$

$E_O = E_i \times N$

$Z_O = Z_i \times N^2$

$P = E_i I_i = E_O I_O$

$\dfrac{E_P}{E_S} = \dfrac{N_P}{N_S}$

$\dfrac{I_S}{I_P} = \dfrac{N_P}{N_S}$

where N_P = number of turns in primary winding

N_S = number of turns in secondary winding

Figure 1-10. Transformers.

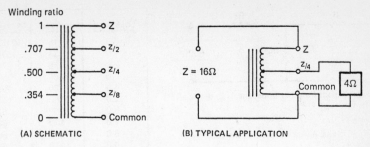

(A) SCHEMATIC   **(B) TYPICAL APPLICATION**

Figure 1-11. The autotransformer.

shows details of the L-pad. These devices are almost always constructed by the user for the specific job at hand. Adjustable wire-wound resistors of adequate power rating should be used.

The L-pad is shown with design data for both 8-ohms and 16-ohms. For most applications, the L-pad is the easiest to implement, and it allows the loudspeaker to look back into a lower resistance than does the T-pad for the same amount of attenuation. The T-pad, shown in Figure 1-13 is best used when both source and load impedances must be carefully matched. This will typically be the case when pads are to be inserted between devices having 600-ohm input and output impedances. Figure 1-13 shows how these pads may be constructed to operate at any impedance.

WIRE LOSSES

It is very important to account for wire losses in all aspects of sound reinforcement, and the cost of larger gauge wire must be justified through value analysis. For example, the calculations for required power in a given system may indicate that peak power capability of 1000 watts is required. A loss of 1 dB in the wiring to the loudspeaker effectively leaves only 800 watts available to the loudspeaker. Also,

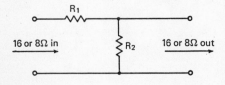

dB LOSS	8Ω R_1	8Ω R_2	16Ω R_1	16Ω R_2
0.5	.4	135.0	.8	270.0
1.0	.8	65.5	1.7	131.0
2.0	1.6	31.0	3.3	62.0
3.0	2.4	19.4	4.7	38.8
4.0	3.0	13.7	5.9	27.4
5.0	3.5	10.3	7.0	20.6
6.0	4.0	8.0	8.0	16.1
7.0	4.5	6.5	8.9	12.9
8.0	4.8	5.3	9.6	10.6
9.0	5.1	4.4	10.3	8.8
10.0	5.5	3.7	10.9	7.4

Figure 1-12. L-pads for adjusting relative drive levels.

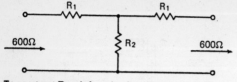

LOSS IN dB	R₁	R₂
0.5	18	10,000
1.0	33	5100
2.0	68	2700
3.0	100	1600
4.0	130	1200
5.0	160	1000
6.0	200	820
7.0	220	680
8.0	240	560
9.0	170	470
10.0	300	430
12.0	360	330
14.0	390	240
16.0	430	200
18.0	470	150
20.0	510	120

To construct T-pads for other impedances, Z, divide all R_1 and R_2 values by $600/Z$.

For example

if Z = 150 ohms, divide all values in the chart by $600/150 = 4$.

Figure 1-13. T-pads for adjusting relative drive levels.

excessive line loss will result in deterioration of loudspeaker response. In general, good engineering practice calls for wire losses to be held to no more than 0.5 dB.

Figure 1-14A presents a table of wire losses for the gauges commonly used in sound reinforcement, and an example of wire loss calculation is shown at B. Here, a load of 8 ohms is intended to be powered with 8 watts. The loss at the load is the result of both line loss and a loss due to the impedance mismatch at the load, which is caused by the added resistance in the line.

Under these conditions, the level loss in the load due to line resistance is:

$$\text{Loss, dB} = 20 \log \left[\frac{R_1}{R_L + 2R_1} \right] \qquad (1\text{-}26)$$

where R_1 is the resistance in each of the two wire runs to the load.

AMERICAN WIRE GAUGE (AWG)	RESISTANCE PER SINGLE RUN, 300 METERS (1000 FEET) OF COPPER (IN OHMS)
5	.3
6	.4
7	.5
8	.6
9	.8
10	1.0
11	1.2
12	1.6
13	2.0
14	2.5
15	3.2
16	4.0
17	5.0
18	6.3
19	8.0
20	10.0

NOTE:

Paralleling two identical gauges reduces effective gauge by 3.

EXAMPLE:

Find the power loss at an 8Ω load due to a 50 meter run of AWG #14 wire.

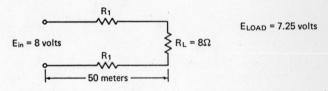

E_{LOAD} = 7.25 volts

$$R = \left(\frac{50}{300}\right) \times 2.5 = 0.416\Omega$$

$$E_{LOAD} = \frac{8}{8 + (2 \times .416)} \times 8 = 7.25 \text{ volts}$$

$$\text{Power in load} = \frac{(7.25)^2}{8} = 6.56 \text{ watts}$$

$$\text{dB loss} = 10 \log\left(\frac{6.56}{8}\right) = .86 \text{ dB}$$

Figure 1-14. Wire losses.

CHAPTER 1:

Recommended Reading:

1. D. and C. Davis; *Sound System Engineering,* Howard W. Sams & Co., Indianapolis (1987).
2. J. Eargle, *Handbook of Recording Engineering,* Van Nostrand Reinhold, New York (1986).
3. J. Eargle and G. Augspurger, *Sound System Design Reference Manual,* JBL Incorporated, Northridge, CA (1986).
4. H. Tremaine, *Audio Cyclopedia,* Howard W. Sams & Co., Indianapolis (1969).
5. G. Ballou, *Handbook for Sound Engineers,* Howard W. Sams & Co., Indianapolis (1987).

Acoustical Fundamentals

INTRODUCTION

In this chapter we present the fundamentals of physical acoustics with special emphasis on the behavior of sound in the indoor environment. The major topics to be covered are: spectra of complex waves, velocity, wavelength and frequency of sound, refraction and diffraction of sound, inverse square law, addition of acoustic signals, line and plane sources, directivity of sound sources and receivers, the absorption, reflection, and reverberation of sound indoors, effects of humidity, and normal mode structure in rectangular rooms.

SPECTRA OF COMPLEX WAVES

In the first chapter, we referred to the sine wave as the fundamental component of periodic waveforms. Sound waveforms exist as pressure maxima and minima in the atmosphere, and they are usually complex in nature. Periodic waveforms are made up of harmonically related sine waves. In Figure 2-1A, we show four sine waves in the frequency relationship of f, $2f$, $3f$, and $4f$. Note that their summation begins to resemble what is called a "sawtooth" waveform. If we were to continue with a very large number of integrally related sine wave components, our resulting waveform would in fact look like a sawtooth wave, as shown at B.

The "square wave" shown at C consists of only odd harmonics: f, $3f$, $5f$, $7f$, and so on.

Noise waveforms are random in nature and are aperiodic, containing all frequencies within a given spectral envelope. *White noise* (shown at D) is so-called because, like white light, it contains all frequencies in the spectrum equally. It contains equal energy per-cycle. *Pink noise* (shown at E) exhibits a rolled off spectrum, just as the color pink does, and it contains equal energy per-octave or fixed portion of an octave. We will deal with pink noise to a great extent in later chapters.

VELOCITY, WAVELENGTH, AND FREQUENCY OF SOUND

Under normal atmospheric conditions, the velocity of sound, c, is given by the following equation:

$$c = 331.4 + 0.607T \text{ meters/sec,} \tag{2-1}$$

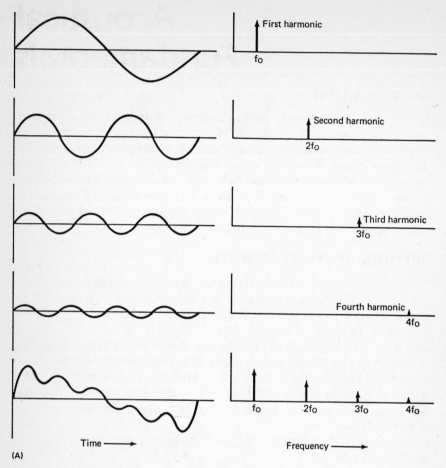

Figure 2-1. The spectra of complex waves.

where T is the temperature in degrees celsius. In English units, the corresponding equation is:

$$c = 1052 + 1.106T \text{ feet/sec}, \tag{2-2}$$

where T is the temperature in degrees fahrenheit.

These equations give the velocity of sound fairly accurately over the range of temperatures that will normally be encountered in sound reinforcement, but for most purposes we can simply ignore the temperature dependence and fix the velocity at a normal temperature of 21 degrees celsius (70 degrees fahrenheit):

$$c = 344 \text{ meters/sec, and} \tag{2-3}$$
$$c = 1130 \text{ feet/sec} \tag{2-4}$$

20

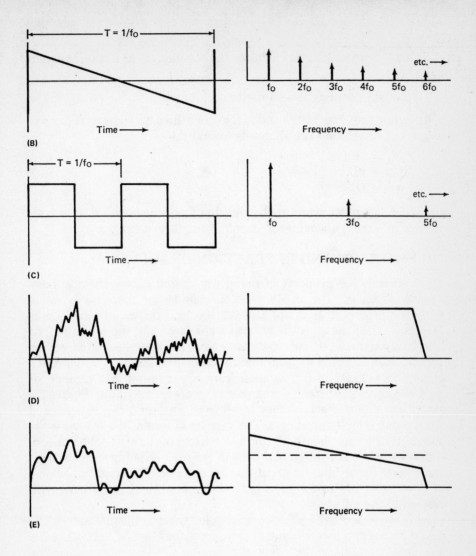

Audible sound covers the spectrum from approximately 20 Hz to 20 kHz. The relationship between the frequency of sound and its wavelength is given by the equation:

$$\text{Wavelength } (\lambda) = \text{velocity/frequency} \tag{2-5}$$

The Greek lambda (λ) is universally used to represent wavelength. As an example, we calculate the wavelength of a 10-kHz signal in air:

$$\lambda = 344/10{,}000 = .0344 \text{ meter } (1\tfrac{1}{3}'')$$

As another example, we calculate the wavelength of a 100-Hz signal in air:

$$\lambda = 344/100 = 3.44 \text{ meters (11.3 feet)}$$

Knowing any two of the following quantities, frequency (f), wavelength (λ), or velocity (v), allows us to find the third:

$$\lambda = v/f \tag{2-6}$$
$$f = v/\lambda \tag{2-7}$$
$$v = f\lambda \tag{2-8}$$

Figure 2-2 presents data showing the frequency spectra of a wide variety of sounds encountered in sound reinforcement work.

DIFFRACTION AND REFRACTION OF SOUND

Diffraction is the property of sound waves that allows them to bend around obstacles. The smaller the obstacle is, the more easily sound will bend around it. If an obstacle is large, or if the sound wavelengths are short, then the object may "cast a shadow," and the sound will be effectively reduced. Figure 2-3 shows several examples of diffraction.

Refraction is the change in the direction of sound as it undergoes a change in velocity of propagation from one medium into another, or as it encounters substantial changes in temperature in air. Figure 2-4 shows refraction effects as they result from temperature gradients and from wind velocity gradients. The velocity of sound in a breeze is the sum of its velocity in still air and the velocity of the air itself. The condition shown at A is often observed in the early evening, when the air has cooled down but the ground is still warm. The condition shown at B often occurs in the morning, when the ground is still cool with respect to the air above.

At C, we observe the effect of wind velocity along the line connecting sound source and listener, and at D we show the effect of a cross breeze. The cross breeze has the effect of changing the axis along which the observer hears the loudspeaker system.

As a general rule, outdoor environment effects are small enough to be ignored. However, very large rock festivals, occurring as they do in the heat of summer, are often affected by these phenomena.

SOUND IN A FREE FIELD: INVERSE SQUARE LAW

A free field is one relatively free of obstacles and reflections of sound. The level produced by a point source located in a free field is observed to fall off at the rate of 6 dB per doubling of distance. The reason for this

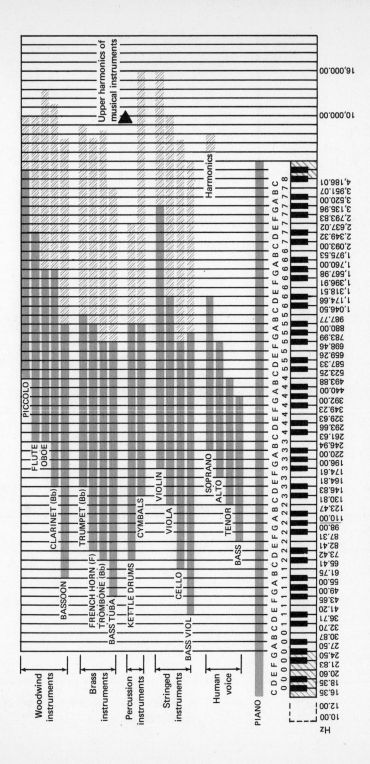

Figure 2-2. Frequency ranges of musical instruments and the human voice.

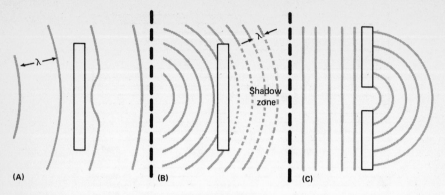

Figure 2-3. Examples of sound diffraction.

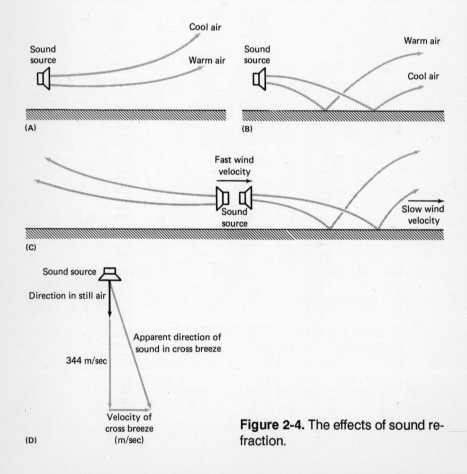

Figure 2-4. The effects of sound refraction.

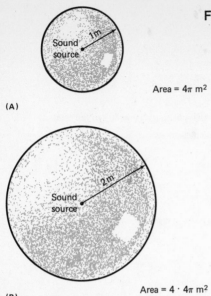

Figure 2-5. The inverse-square law

Area = 4π m^2

(A)

Area = $4 \cdot 4\pi$ m^2

(B)

is shown in Figure 2-5. At A, a sphere with a radius of one meter surrounds a point source, and the sound pressure is noted at the surface. At B, the same sound source is surrounded by a sphere with a radius of two meters. Since the area of a sphere is proportional to the square of its radius, the total area of the sphere at B will be *four times* that of the sphere shown at A.

The power per unit area passing through the sphere at B will then be *one-fourth* that observed at A, and the pressure level will be 6 dB lower, corresponding to a power ratio of four-to-one.

We can state this relationship by the following equation:

Level difference $= 10 \log [d_1/d_0]^2 = 20 \log [d_1/d_0]$ $\hspace{2em}$ (2-9)

where d_0 is the reference distance and d_1 the distance at which we are observing the sound pressure level.

Sound pressure level (SPL) is usually measured with an instrument called a sound level meter. The reference for zero dB is near the threshold of hearing and is equal to 2×10^{-5} pascal (.0002 dynes/cm^2). The term L_p is also used to indicate sound pressure level relative to 2×10^{-5} pascal.

SPL is analogous to potential in electrical circuits and is given by the following equation:

SPL $= 20 \log (P_1/P_0)$, $\hspace{4em}$ (2-10)

where P_0 is the reference level stated above.

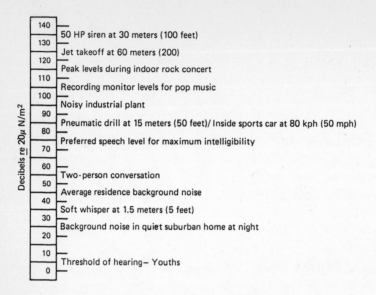

Figure 2-6. Typical levels of common sound sources that are A-weighted.

Sound pressure levels of common sources are indicated in Figure 2-6.

A convenient nomograph for determining inverse square losses is given in Figure 2-7. As an example of how to use it, a sound source is observed to produce an SPL of 92 dB at a distance of 6 meters. What will be the observed SPL at 18 meters? Above 6 read 15.5 dB, and above 18 read approximately 25. Taking the difference, 25 − 15.5 = 9.5 dB.

SUMMING ACOUSTIC LEVELS

We cannot merely add levels in dB directly to get their sum, and the nomograph of Figure 2-8 allows us to combine them by inspection. First, we take the difference between the two levels and locate the number along the top of the nomograph. Then we look at the corresponding number along the lower part of the nomograph and add that number to the higher of the two original levels.

We will work an example: Find the sum of the acoustical levels of 93 dB and 100 dB. First, we take their difference, 7 dB, and locate that at the top of the nomograph. The number N corresponding to this is 0.8, and the resultant level will then be 100 + 0.8 = 100.8 dB SPL.

Note that when two levels differ by more than 10 dB, the combination of the two is nearly insignificant compared to the louder of the two original levels.

This nomograph is not limited to determining the sums of acoustical levels. It can be used for all power summations when those powers are

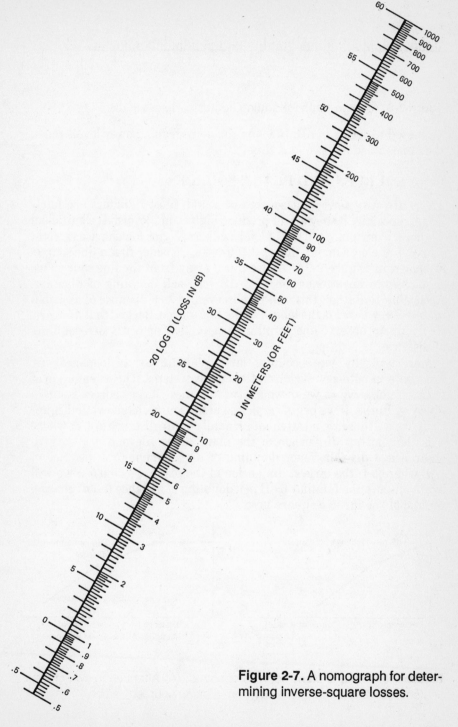

Figure 2-7. A nomograph for determining inverse-square losses.

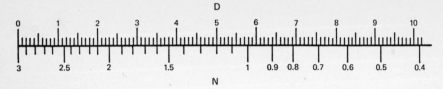

Figure 2-8. A nomograph for adding acoustical levels in dB.

expressed as levels in dB. In using the nomograph, power levels must, of course, be added two at a time.

LINE SOURCES AND PLANE SOURCES

There are very few point sources of sound. Most radiators are fairly large, especially if they are to produce high sound levels. At a sufficient distance, even a large array of loudspeakers begins to behave as a point source, but closer in, it behaves differently. Consider first a line source, as shown in Figure 2-9A. As we move away from the line source, we first observe an attenuation of 3 dB for each doubling of distance. According to Rathe[1], this will be observed up to a distance of approximately l/π, where l is the length of the line source. Beyond that distance, we begin to observe the familiar attenuation of 6 dB per-doubling of distance.

More typically, we encounter large plane arrays of loudspeakers, especially in outdoor music reinforcement systems. The attenuation of sound we observe as we move away from these large sources is again given by Rathe. If we begin very close to the source, as shown in Figure 2-9B, we will observe no attenuation at all out to a distance of a/π, where a is the shortest dimension of the plane array. Beyond a/π we will observe a 3 dB falloff per doubling of distance out to a distance of b/π, where b is the largest dimension of the plane. Beyond b/π we will again observe the familiar 6 dB per-doubling of distance falloff characteristic of the inverse square law.

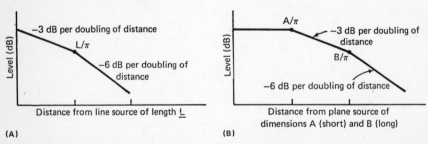

(A)

(B)

Figure 2-9. Line and plane sources of sound. **(A)** Attenuation away from a line source. **(B)** Attenuation away from a plane source.

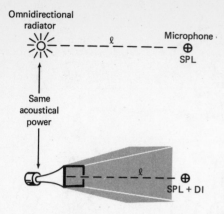

Figure 2-10. The definition of directivity index (DI).

The complexities observed here mean that it is difficult to estimate just how loud a large loudspeaker array will be at some distance without a good working knowledge of how the components interact. We will discuss this in greater detail in a later chapter.

DIRECTIVITY OF SOUND SOURCES AND RECEIVERS

When we choose microphones or loudspeakers for sound reinforcement, we usually aim them at their respective targets. A directional microphone will have its major axis pointed at a talker, while the major axis of the loudspeaker will be pointed in the direction of the audience. A very convenient measure of how directional a device is can be found by its directivity index (DI). It is defined as shown in Figure 2-10. Thus, if a given loudspeaker has a DI of 13 dB, it will, along its major axis, produce a level some 13 dB louder than would the same amount of acoustical power radiated omnidirectionally from the same point.

A related quantity is the directivity factor of the device. The symbol Q is generally used to represent directivity factor despite the possibility of its being confused with the Q of a resonant circuit. The two are not related. Other symbols used for directivity factor are DF and R_θ. Q is related to DI as follows:

$$Q = 10^{\frac{DI}{10}}, \text{ and} \tag{2-11}$$
$$DI = 10 \log (Q) \tag{2-12}$$

Figure 2-11 shows the directional characteristics over a wavelength range of 24-to-1 for a circular piston mounted in a wall, corresponding, for example, to a typical single-cone 300 mm (12-inch) loudspeaker covering the frequency range from about 250 Hz up to about 6 kHz. Both Q and DI are indicated. Note that when the loudspeaker is small compared to the wavelength, it is nearly omnidirectional, but with

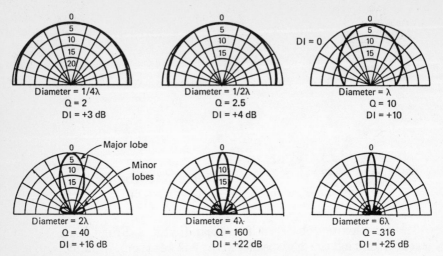

Figure 2-11. Directional characteristics of a circular piston source mounted in an infinite baffle as a function of diameter and λ.

rising frequency, the loudspeaker begins to beam, and at the highest frequencies we see the formation of minor lobes in the response. The upper right illustration gives an indication of the off-axis angle at which the DI is zero. The reader must be aware that both DI and Q are functions of the angle along which we observe a loudspeaker. The published values of DI and Q are always those taken along the major sound output axis of the device.

A more useful presentation of directional information on loudspeakers is beamwidth. Generally, in professional sound work, the –6 dB angles are plotted, as shown in Figure 2-12, along with a plot of DI and Q as a function of frequency. The information presented in this figure will give the user of a device just about all the information he will need in system layout.

Where loudspeakers exhibit fairly sharp cut-off outside the –6 dB included angle, we can get a reasonable approximation of DI and Q through the use of an equation developed by Molloy[2], as shown in

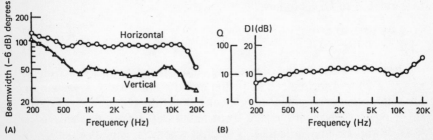

Figure 2-12. Beamwidth and directivity vs. frequency.

30

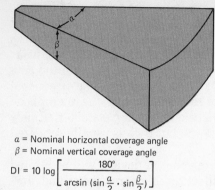

Figure 2-13. The illustration of Molloy's equation.

a = Nominal horizontal coverage angle
β = Nominal vertical coverage angle

$$DI = 10 \log \left[\frac{180°}{\arcsin \left(\sin \frac{a}{2} \cdot \sin \frac{\beta}{2} \right)} \right]$$

$$Q = \frac{180°}{\arcsin \left(\sin \frac{a}{2} \cdot \sin \frac{\beta}{2} \right)}$$

Figure 2-13. For most applications involving high-frequency horns, Molloy's equation will give good estimates of DI and Q.

In contrast with loudspeakers, microphones have relatively simple pickup patterns. The three patterns most often encountered are shown in Figure 2-14. Values of DI and Q along the major axes are indicated. It should be noted that microphones will, at high frequencies, begin to beam just as loudspeakers will. Because of the small size of the diaphragms, however, this effect usually takes place well up in the frequency range and, as a result, may not be a problem.

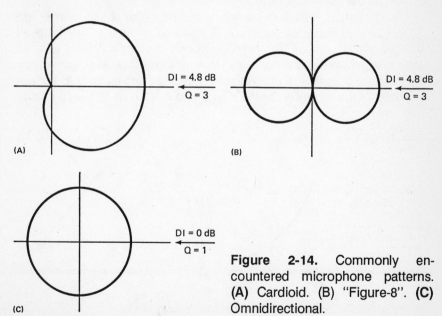

(A)

DI = 4.8 dB
Q = 3

(B)

DI = 4.8 dB
Q = 3

(C)

DI = 0 dB
Q = 1

Figure 2-14. Commonly encountered microphone patterns. **(A)** Cardioid. **(B)** "Figure-8". **(C)** Omnidirectional.

The graphs shown in Figure 2-14 are known as *polar plots*. In these graphs, the signal amplitude is plotted as a function of angle about the microphone. The amplitude can be expressed as pressure response, as shown here, or it can be expressed as level, in dB. The utility of polar plots is that they give us a clear picture of response through the 360-degree arc around a device. We will always see microphone directional response shown in this manner, since sound may enter the microphone from just about any direction. Loudspeaker polar plots are routinely made by manufacturers of loudspeakers, but data is usually replotted in the format shown in Figure 2-12.

Our coverage of the directional properties of devices will be limited in this chapter. Later chapters will cover high-frequency and low-frequency components in detail, and we will expand our discussion of their directional properties in those chapters.

GROWTH AND DECAY OF INDOOR SOUND FIELDS

If we turn a sound source on and off in a room and look at the growth and decay of the sound field, we will see something like that shown in Figure 2-15. There is a growth period which is exactly equal to the decay period, as shown at A. However, if we plot the level in dB, as shown at B, the growth period appears to be much shorter than the decay period. This is in fact the way we hear it, inasmuch as the sensation of loudness is proportional to the logarithm of sound power, as we will see in the following chapter.

What is actually happening when we turn the sound source on? At first, sound radiates out from the source, eventually striking all boundaries of the room. The process quickly reaches a state of equilibrium in which sound is absorbed at the same rate it is being generated. The average path length of sound as it reflects from surface to surface is known as the *mean free path* (MFP), and it is given by the following equation:

$$MFP = 4V/S, \tag{2-13}$$

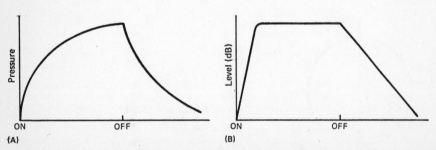

Figure 2-15. Growth and decay of indoor sound fields.

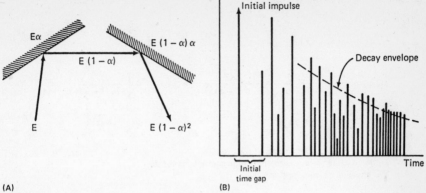

Figure 2-16. Sound reflections in a room.

where V is the room volume and S the surface area. The equation is applicable to any system of units.

Each surface of the room boundaries has an *absorption coefficient*, α, which determines the amount of sound absorbed and the amount reflected. For example, a material with an absorption coefficient of .37 will absorb 37% of the sound incident on it and reflect the remaining 63%. As shown in Figure 2-16A, successive reflections of sound bring into play all the room surfaces, and we can speak, in a statistical sense, of an *average absorption coefficient* in the room. The symbol for average absorption coefficient is $\bar{\alpha}$.

At Figure 2-16B, we show the discrete nature of typical room reflections as observed at a fixed point in the room. Individually, the reflections are almost of random amplitude and spacing, but their overall envelope conforms to an exponential decay. The initial time gap, indicated in the figure, is an important factor in determining certain subjective attributes of speech and music performance in the room.

Reverberation time (RT$_{60}$) is that time required for sound to decay 60 dB in level after the sound source has been turned off. Many years ago, Sabine[3] derived the following equation for reverberation time:

$$RT_{60} = \frac{.16V}{S\bar{\alpha}} \qquad (2\text{-}14)$$

where $S\bar{\alpha}$ = surface area times absorption coefficient, measured in square meters.

In English units, the equation is:

$$RT_{60} = \frac{.05V}{S\bar{\alpha}} \qquad (2\text{-}15)$$

where $S\bar{\alpha}$ is measured in square feet.

33

EQUATION	METRIC (SI) UNITS	ENGLISH UNITS

Norris-Eyring:

$$RT_{60} = \frac{.16\ V}{-S\ \ln(1-\bar{\alpha})} \qquad (2-16) \qquad RT_{60} = \frac{.05\ V}{-S\ \ln(1-\bar{\alpha})} \qquad (2-17)$$

Fitzroy:

$$RT_{60} = \frac{.16\,V}{S^2}\left[\frac{2XY}{-\ln(1-\alpha_{XY})} + \frac{2XZ}{-\ln(1-\alpha_{XZ})}\right.$$
$$\left. + \frac{2YZ}{-\ln(1-\alpha_{YZ})}\right] \qquad (2-18)$$

$$RT_{60} = \frac{.05\,V}{S^2}\left[\frac{2XY}{-\ln(1-\alpha_{XY})} + \frac{2XZ}{-\ln(1-\alpha_{XZ})}\right.$$
$$\left. + \frac{2YZ}{-\ln(1-\alpha_{YZ})}\right] \qquad (2-19)$$

NOTES:

1. X, Y and Z are the basic dimensions of the room.
2. α_{XY}, α_{XZ} and α_{YZ} are the averaged absorption coefficients for the X-Y, X-Z and Y-Z surfaces, respectively.

Figure 2-17. Norris-Eyring and Fitzroy reverberation-time equations.

Later studies have refined Sabine's work, and the equations shown in Figure 2-17 are widely used today. The Norris-Eyring equation is appropriate where absorption is evenly distributed throughout the space, while the Fitzroy equation would be used in a room where the absorption was not evenly distributed. Sabine's work was very important to the development of architectural acoustics, and his name is given to the unit of absorption, the *sabin*, which we will now define.

Let us assume that we have an area of 100 square meters with an absorption coefficient of 0.6. Then:

Absorption = $S\alpha$ = (100) (0.6) = 60m².

What this means is that the surface acts as though it consists of 60 square meters of totally absorptive surface. Obviously, there are both metric (SI) and English sabins, square meters or square feet, depending on the units we are working in.

Figure 2-18 presents a table of absorption coefficients for various materials and treatments. Note that there is considerable variation with frequency, and this means that reverberation time will vary with frequency.

The calculation of reverberation time, as opposed to the measurement of it, is a rather tedious procedure. First, we calculate the volume and surface area. Then, we use the following equation to arrive at the average absorption coefficient:

$$\bar{\alpha} = \frac{S_1\alpha_1 + S_2\alpha_2 + \ldots + S_n\alpha_n}{S_1 + S_2 + \ldots + S_n} \qquad (2-20)$$

In this equation, $S_{1\ldots n}$ and $\alpha_{1\ldots n}$ represent the individual areas of room surface with their respective absorption coefficients.

MATERIAL	125	250	500	1KHz	2KHz	4KHz
Brick wall (18" thick, unpainted)	.02	.03	.03	.04	.05	.07
Brick wall (18" thick, painted)	.01	.01	.02	.02	.02	.02
Interior plaster on metal lath	.02	.03	.04	.06	.04	.03
Poured concrete	.01	.01	.015	.02	.02	.03
Pine flooring	.09	.11	.10	.08	.08	.10
Carpeting with pad	.10	.25	.60	.70	.70	.70
Drapes (cotton, 2x fullness)	.07	.30	.50	.80	.65	.50
Drapes (velour, 2x fullness)	.15	.35	.55	.75	.70	.65
Acoustic tile ($5/8$", #1 mount)	.15	.35	.50	.70	.70	.65
Acoustic tile ($5/8$", #2 mount)	.25	.40	.55	.70	.70	.65
Acoustic tile ($5/8$", #7 mount)	.50	.55	.60	.75	.70	.65
Tectum panels (1", #2 mount)	.08	.15	.30	.55	.60	.65
Tectum panels (1", #7 mount)	.35	.40	.45	.35	.50	.65
Plywood paneling ($1/8$", 2" air space)	.30	.25	.20	.10	.08	.07
Plywood cylinders (2 layers, $1/8$")	.35	.30	.25	.20	.20	.18
Perforated transite (W/pad, #7 mount)	.90	.90	.90	.95	.60	.45
Occupied audience area	.50	.70	.80	.95	.90	.85
Upholstered theater seats on hard floor	.45	.70	.80	.90	.80	.70

NOTES:

#1 mount = Cemented directly to plaster or concrete
#2 mount = Fastened to nominal 1" thick furring strips
#7 mount = Suspended ceiling with 16" air space above

Figure 2-18. Absorption coefficients.

Working an example, we calculate the reverberation time at 500 Hz for the room described below:

Dimensions: length = 20 meters
width = 12 meters
height = 7 meters
volume = 1680 meters^3

Surface details:

Materials:	α:	*Dimensions:*	*Area:*	*Sα:*
carpet (floor)	.6	20 x 12	240 m²	144 m²
plywood (ceiling)	.2	20 x 12	240 m²	48 m²
drapes (wall)	.55	12 x 7	84 m²	46.2 m²
plaster (wall)	.04	20 x 7	140 m²	5.6 m²
plaster (wall)	.04	20 x 7	140 m²	5.6 m²
acoustic tile (wall)	.5	12 x 7	84 m²	42 m²
			928 m²	291.4 m²

Average absorption coefficient, $\bar{\alpha} = (\text{total } S\alpha)/S = \dfrac{291.4}{928} = .31$

And finally, $RT_{60} = \dfrac{.16\ (1680)}{-\ 928\ \ln\ (1 - .31)} = .78$ sec

Since this room does not exhibit uniform distribution of absorption, it would appear to be a good candidate for the Fitzroy reverberation time equation. We will calculate the reverberation time using that equation:

$$RT_{60} = \frac{.16\,(1680)}{(928)^2} \left[\frac{2 \times 20 \times 12}{-\ln(1-.4)} + \frac{2 \times 12 \times 7}{-\ln(1-.52)} + \frac{2 \times 20 \times 7}{-\ln(1-.04)} \right]$$

$$RT_{60} = \frac{269}{861184} \left[\frac{480}{.51} + \frac{168}{.73} + \frac{280}{.04} \right]$$

$$RT_{60} = .0003\,(941 + 230 + 7000)$$

$$RT_{60} = .0003\,(8171) = 2.4 \text{ seconds}$$

The Fitzroy equation gives a value of reverberation time about three times that of the Norris-Eyring equation and is probably a more accurate estimate of what will actually take place in the room. Note that this room has a pair of opposite walls whose absorption coefficient is considerably lower than the others in the room. The implication here is that sound will reverberate between these surfaces for some time after it has died down in other directions in the room. We recognize this in many rooms as a "flutter echo" between parallel reflective surfaces, and the Fitzroy equation takes such conditions into consideration.

There is considerable variation in reverberation time from room to room, but larger rooms tend to have longer reverberation times than small ones. Spaces used for speech communication are generally designed to have shorter reverberation times than those used for music performance. Optimum values of reverberation time as a function of room volume and intended purposes are given in Figure 2-19.

ATTENUATION OF SOUND WITH DISTANCE INDOORS

If we move away from a sound source in a reverberant room, we will observe that the level falls off as shown in Figure 2-20. There is a region where the inverse square law holds, followed by a transition into a region where the level remains constant. The reverberant level is in fact constant throughout the room if there is good diffusion and if the average absorption coefficient in the space is less than about 0.2. The distance from the source where the direct and reverberant fields are equal is called the critical distance (D_c), and at this distance the direct-to-reverberant ratio is unity, or zero, if expressed in dB.

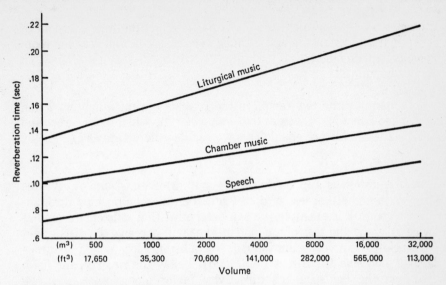

Figure 2-19. Optimum reverberation times as a function of volume and purpose.

The equation describing the curve shown in Figure 2-20 is given as:

$$\text{Pressure (pascals)} = \left[W\rho_0 c \left(\frac{4}{R} + \frac{Q}{4\pi r^2} \right) \right]^{\frac{1}{2}} \qquad (2\text{-}21)$$

Expressing this equation in dB:

$$\text{SPL} = 10 \log W + 10 \log \rho_0 c + 94 + 10 \log \left(\frac{4}{R} + \frac{Q}{4\pi r^2} \right) \quad (2\text{-}22)$$

In these equations, W is the acoustical power output of the loudspeaker, $\rho_0 c$ is the acoustical impedance of air (415 N·sec/m²), Q is the directivity

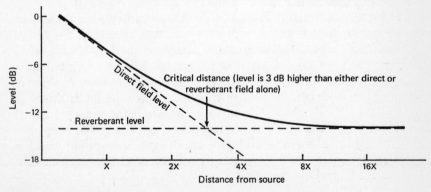

Figure 2-20. Sound attenuation with distance indoors.

factor of the loudspeaker along the direction of the measurement of level, r is the distance away from the loudspeaker, and R is the room constant, given below:

$$R = S\bar{\alpha}/(1 - \bar{\alpha}) \qquad (2\text{-}23)$$

If we equate the two terms in the brackets in the equations given above and solve for r, we get that value of r for which both direct and reverberant fields are equal. This we earlier referred to as D_c.

$$D_c = .14\sqrt{QR} \qquad (2\text{-}24)$$

This equation is applicable to any system of units, and it gives us almost immediate access to direct-reverberant relationships throughout the room. For example, we can determine that a listener seated at twice D_c along the axis of the loudspeaker will perceive the direct field 6 dB below the reverberant field, or that a listener located at four-times D_c will perceive the direct field 12 dB below the reverberant field. Knowing the loudspeaker's directivity factor along other axes, we can determine direct-reverberant relationships in those directions as well.

REVERBERANT FIELD CALCULATIONS

While a knowledge of critical distance allows us to determine direct-to-reverberant ratios throughout a room, it is usually easier to solve for the reverberant level directly and compare it with the calculated inverse square determination of the direct field level. The symbol SPL_{rev} used to indicate the level of the reverberant field in dB, and the equation which gives the reverberant level is:

$$SPL_{rev} = 10 \log (W/R) + 126 \text{ dB} \qquad (2\text{-}25)$$

where W is the total acoustical power output of the loudspeaker, or other source generating sound in the room.

R as we have defined it is applicable to a room in which absorption is evenly distributed, and it will give accurate results in the above equation if an omnidirectional sound source is used. But note what happens when we take a directional loudspeaker and point it at a highly absorptive surface, as shown in Figure 2-21. In Case A, there is a fairly high reverberant level in the room, since the loudspeaker is not aimed at a highly absorptive surface. In Case B, the same acoustical power is being radiated by the loudspeaker, but the initial reflection from the highly absorptive audience area is much less than was the initial reflection in Case A. Augspurger[4] proposed a new room constant, called R' and defined it as:

$$R' + S\bar{\alpha}/(1 - \alpha_1) \qquad (2\text{-}26)$$

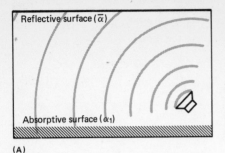

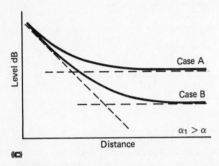

Figure 2-21. A loudspeaker in a room with highly-absorptive material.

where α_1 is the absorption coefficient of the first surface struck by sound from the directional loudspeaker.

Which is the right value of R to use in our equations? Obviously, it depends on the nature of the problem and the type of loudspeaker used. It is clear that for many applications in sound reinforcement, R′ will give more accurate values of the reverberant level in the room.

In Figures 2-22 and 2-23, it is apparent that either R or R' can be used in determining SPL in the room. These sets of graphs will be useful in sound contracting work for determining actual reverberant levels in a room from a knowledge of system efficiency, electrical power input, and room constant.

The graphs of Figure 2-24 will enable the designer to solve for room constant by inspection, and the graphs of Figure 2-25 will give critical distance by inspection.

The graphs of Figure 2-26 relate room volume and reverberation time to the acoustical power required to attain a level of 94 dB-SPL, or one pascal.

Finally, the graphs given in Figure 2-27 relate reverberation time, room constant, and volume.

EFFECTS OF HUMIDITY

The effect of humidity on the absorption of sound in air is a complex one, and the data of Figure 2-28 may be used to calculate the attenuation of sound with distance in addition to that determined by inverse

39

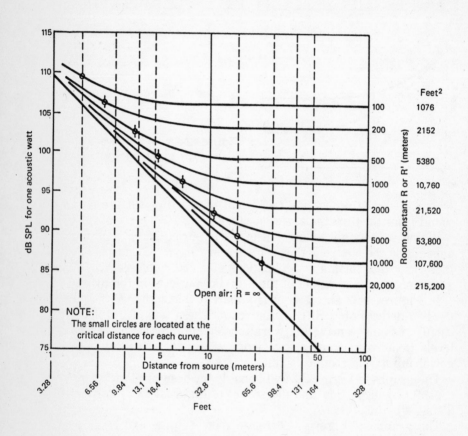

Figure 2-22. SPL (point source radiating one acoustical watt vs. R and distance from source.)

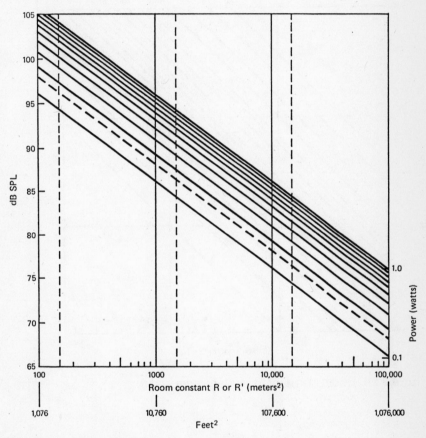

Figure 2-23. Reverberant level vs. R and acoustical power.

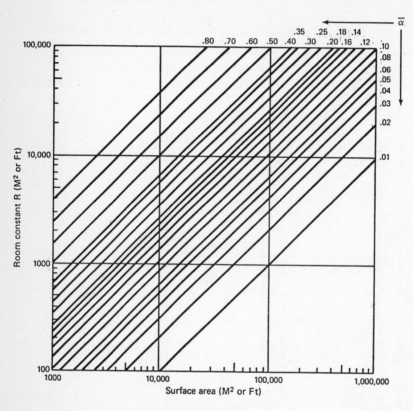

Figure 2-24. Room constant vs. surface area and $\overline{\alpha}$.

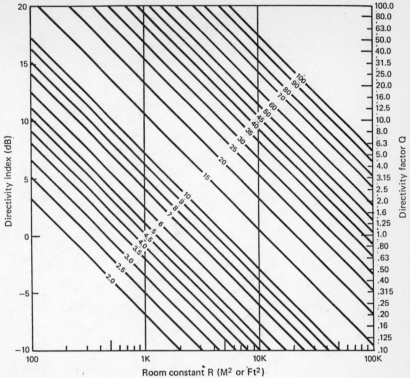

NOTE:
  Diagonal lines: Critical distance (M or Ft)
  Equations: Critical distance = $0.14\sqrt{QR}$.
  Directivity index (dB) = $10 \log Q$.

Equations and chart can be used with English or metric units. To convert chart scales to more convenient values for metric calculations, divide critical distances by 10 and room constants by 100.

Figure 2-25. Critical distance as a function of room constant and DI or Q.

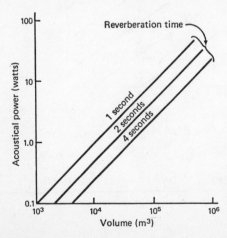

Figure 2-26. The relationship between room volume and reverberation time to the acoustical power required to produce 94 dB SPL.

43

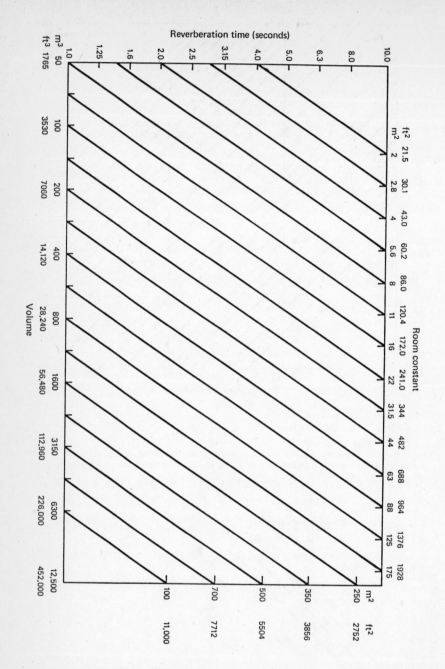

Figure 2-27. Estimation of room constant when volume and T60 are known.

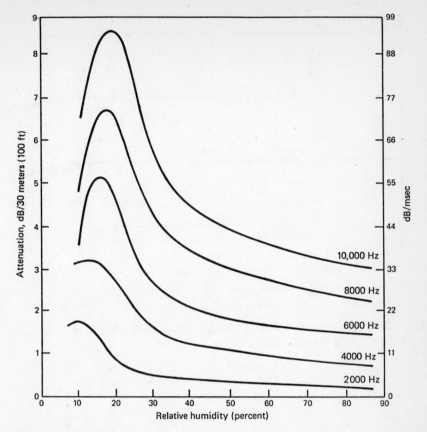

Figure 2-28. The absorption of sound in air vs. frequency and relative humidity.

square relationships. Figure 2-29 shows typical losses with distance and as a function of relative humidity. Note that the drier the air is, the greater the absorption.

Air absorption due to humidity will have the effect of altering reverberation times, as given by the data of Figure 2-30.

QUASI-STEADY-STATE SOUND FIELDS

Our discussion of reverberant sound fields assumed that a uniform field existed throughout the room. While this is certainly true for quite reverberant spaces, most rooms have sufficient absorption so that the reverberant field cannot establish absolute uniformity throughout the room. What we usually observe when we move away from a sound source in most rooms is shown in Figure 2-31. Peutz[5] made many measurements of levels beyond critical distance and

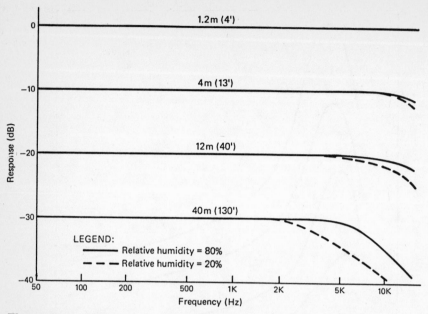

Figure 2-29. Losses due to both inverse-square law falloff and atmospheric absorption at high frequencies. (T = 20° C)

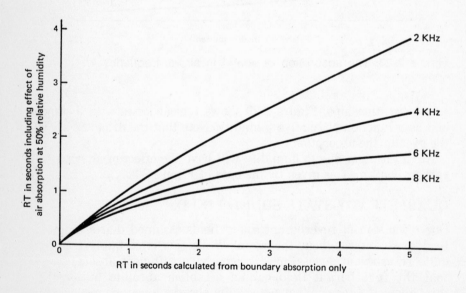

Figure 2-30. The effect of air absorption on calculated reverberation time.

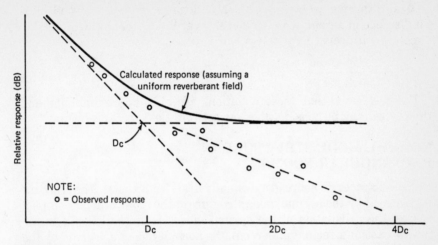

Figure 2-31. Attenuation with distance in a relatively dry acoustical environment.

derived empirical equations that will enable a designer to estimate the slope of the dotted line in the graph. For rooms of fairly regular dimensions, the additional loss in level with doubling of distance beyond critical distance is given by:

$$\Delta = \frac{0.4 \sqrt[6]{V}}{RT_{60}} \qquad (2\text{-}27)$$

In this equation, Δ represents the falloff in level per doubling of distance beyond critical distance, V is the room volume in m^3, and RT_{60} is the reverberation time in seconds.

For rooms with relatively low ceilings, the corresponding equation is:

$$\Delta = \frac{0.4 \sqrt{V}}{hRT_{60}} \qquad (2\text{-}28)$$

In this equation, h is the room height in meters.

In English units these equations are:

$$\Delta = \frac{0.22 \sqrt[6]{V}}{RT_{60}} \qquad (2\text{-}29)$$

and

$$\Delta = \frac{0.22 \sqrt{V}}{hRT_{60}} \qquad (2\text{-}30)$$

47

As an example, we will find the additional attenuation beyond critical distance in a room whose dimensions are 15 meters by 10 meters by 3 meters with reverberation time of 1 second.

$$\Delta = \frac{0.4 \sqrt{450}}{3 \times 1} = 2.8 \text{ dB}$$

Peutz stresses that these equations are only approximations and that there may be considerable variation in field measurements.

NORMAL MODE STRUCTURE IN RECTANGULAR ROOMS

Enclosed spaces do not respond equally to all frequencies. The response is governed by the normal mode structure of the room, and for complex spaces, the calculation of the normal modes is quite difficult. In the simple case of a rectangular room, the normal modes are given by the following equation:

$$f = \frac{c}{2} \sqrt{\left(\frac{n_l}{l}\right)^2 + \left(\frac{n_w}{w}\right)^2 + \left(\frac{n_h}{h}\right)^2} \qquad (2\text{-}31)$$

In this equation, c is the velocity of sound in air, and l, w, and h are, respectively, the length, width, and height of the room. The three values of n in the equation are independent sets of integers. The lowest normal mode of the room is determined when n corresponding to the longest dimension is set at unity and the other two values of n are set at zero.

In the example of Figure 2-32, the first few modes are for a room of dimensions 4 by 6 by 10 meters. The integers in parentheses next to each mode indicate the three values of n in the equation.

Room modes can be troublesome in smaller spaces, where their peaks and dips may extend up into the mid-range. In very large rooms, churches and auditoriums, for example, the mode structure begins very low in frequency, and by the time it reaches the 50 Hz range and upwards, the mode structure becomes quite dense, uniform, and relatively free of coloration. The effect of room modes in smaller rooms can be minimized by adding low frequency absorption to the room.

CHAPTER 2:

References:
1. E. Rathe, "Note on Two Common Problems of Sound Propagation," *J. Sound Vibration.* Vol. 10, pp. 472-479 (1969).
2. C. T. Molloy, "Calculation of the Directivity Index for Various Types of Radiators," *J. Acoustical Soc. Am.,* Vol. 20: 387-405 (1948).
3. W. Sabine, "Architectural Acoustics," *Engineering Record* (1900).
4. G. Augspurger, "More Accurate Calculation of the Room Constant," *J. Audio Eng. Soc.,* Vol. 23, No. 5 (1975).
5. V.M.A. Peutz, "Quasi-steady-state and Decaying Sound Fields," *Ingenieursblad,* Vol. 42, No. 18 (Sept 1973), (in Dutch).

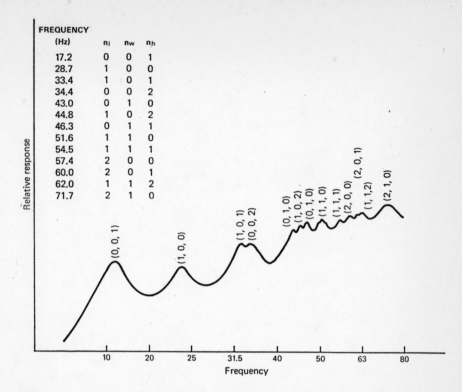

FREQUENCY (Hz)	n_l	n_w	n_h
17.2	0	0	1
28.7	1	0	0
33.4	1	0	1
34.4	0	0	2
43.0	0	1	0
44.8	1	0	2
46.3	0	1	1
51.6	1	1	0
54.5	1	1	1
57.4	2	0	0
60.0	2	0	1
62.0	1	1	2
71.7	2	1	0

Figure 2-32. Low-frequency room modes in a room 6x4x10 meters.

Recommended Reading:

Article:

1. H. Hopkins and N. Stryker, "A Proposed Loudness-efficiency Rating for Loudspeakers and the Determination of System Power Requirements for Enclosures," *Proceedings of the IRE*, March (1948).

Books:

1. L. Beranek, *Acoustics*, McGraw-Hill, New York (1954).
2. L. Beranek, *Music, Acoustics, and Architecture*, John Wiley & Sons, New York (1962).
3. D. and C. Davis, *Sound System Engineering*, Howard W. Sams & Co., Indianapolis (1987).
4. V. Knudsen and C. Harris, *Acoustical Designing in Architecture*, Acoustical Society of America, New York (1978).
5. H. Kuttruff, *Room Acoustics*, Applied Science Publishers, London (1979).
6. M. Rettinger, *Acoustic Design and Noise Control*, Chemical Publishing Co., New York (1977).
7. Anon, *Acoustics Handbook*, Application Note 100, Hewlett-Packard, Palo Alto, CA (1968).

Psychoacoustical Aspects

INTRODUCTION

Psychoacoustics is the study of how the ear-brain combination responds to auditory stimuli. Our discussion will emphasize those aspects of psychoacoustics which are important in understanding how musicians respond to the world of sound around them. Topics to be discussed include: loudness contours and weighting curves, loudness calculations, dependence of loudness on sound duration, masking, critical bandwidth, pitch phenomena, the precedence effect, and subjective aspects of performance spaces.

LOUDNESS CONTOURS

We have all had the experience of hearing a stereo system playing at a normal level, sounding properly balanced throughout the frequency band. When the volume is turned down, the reproduction sounds thin and bass-shy, even though the frequency spectrum has not been altered. What this demonstrates is the well-known fact that the ear is less sensitive to low frequencies at low levels than it is to higher frequencies. Fletcher and Munson studied this phenomena in 1933, and their equal loudness contours are well known. A more recent study was made by Robinson and Dadson[1], and their curves are generally felt to be more reliable than those of Fletcher and Munson. These are shown in Figure 3-1. The curve labeled MAF corresponds to the minimum audible field (MAF).

Examining these curves, we see that at 1 kHz the loudness of a pure tone in *phons* is equal to the SPL. The reference frequency for determining the equal loudness contours was 1 kHz, and in determining these curves, subjects were asked to compare variable tones with fixed levels at 1 kHz and adjust those tones so they were equal in loudness to the 1 kHz reference tones. Observe that the 20-phon curve at 100 Hz corresponds to about 37 dB-SPL. What this means is that a 100 Hz tone at 37 dB SPL will sound as loud as a 1 kHz tone at 20 dB-SPL.

At the lowest levels there is the threshold of audibility, or minimum audible field, and in this range the ear can detect air particle velocity as small as 2×10^{-6} cm/sec. At the highest levels, in the 120 phon range,

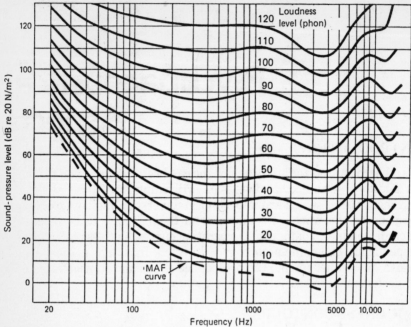

Figure 3-1. Equal loudness contours.

we reach the threshold of feeling, and the corresponding air particle velocities are as great as 2 cm/sec.

WEIGHTING CURVES

Taking into account the ear's reduced sensitivity to low frequencies at low levels, weighting curves have been developed for use with sound level meters. Figure 3-2A shows a modern high-quality sound level meter, and at B we show the various weighting curves incorporated into it. Note that the "A" and "B" weighting curves are approximately the inverse of the 40 and 70-phon curves shown in Figure 3-1. When these weighting curves are used in their respective ranges, the sound level meter will read loudness levels more in agreement with our subjective judgements.

While the data shown in Figure 3-1 tells us a great deal about the relative loudness of different frequencies at different levels, it does not tell us whether a given sound is twice as loud, or half as loud as another one. Based on psychoacoustical studies, a sound must be raised some 10 dB in level in order to sound "twice as loud." This relationship points out one of the difficulties in high-level sound reinforcement. To make a system play subjectively "twice as loud" not only may call for more in the way of loudspeakers, but will require ten times the power input.

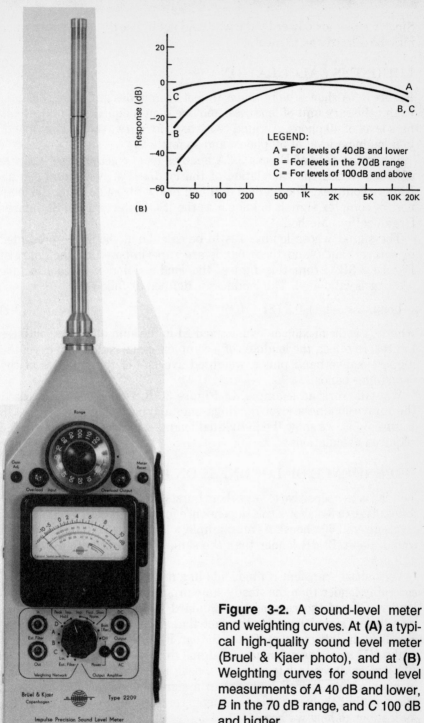

(B)

Figure 3-2. A sound-level meter and weighting curves. At **(A)** a typical high-quality sound level meter (Bruel & Kjaer photo), and at **(B)** Weighting curves for sound level measurments of *A* 40 dB and lower, *B* in the 70 dB range, and *C* 100 dB and higher.

Since we pay for power by the watt and not by the dB, the requirement may be a hard one to meet.

LOUDNESS CALCULATION

Figure 3-3A shows the relationship between phons and sones. The *sone* is an arbitrary unit of loudness, and one sone is equivalent to the loudness level of 40 phons. A sound twice as loud, or two sones, corresponds to a loudness level of 50 phons, and so forth.

While the data of Figure 3-3A enables us to calculate the loudness of single tones, the calculation of the loudness of wide-band signals is more complex. The method outlined below for calculating the loudness of complex signals is known as the ISO (International Standards Organization) Method A:

The signal whose loudness is to be calculated should be presented in octave bands, and these bands are superimposed on the chart of Figure 3-3B. From this figure, the loudness index of each octave reading is tabulated. The loudness is defined as follows:

$$\text{Loudness} = I_m + 0.3\,(\Sigma I - I_m) \tag{3-1}$$

where I_m is the maximum loudness and ΣI is the sum of all the loudness indices. In effect, the loudness of a complex signal is the loudness of the highest octave band plus a weighted average of the loudness of the remaining bands.

We will work an example. In Figure 3-3B, we have superimposed the octave loudness level readings onto a typical noise spectrum. In Figure 3-3C, we show the individual loudness indices and perform the required calculations.

DEPENDENCE OF LOUDNESS ON DURATION

The ear is less sensitive to very short bursts of tone than it is to sustained tones. Figure 3-4 shows this dependence for tones in the 1 kHz range at levels around 60 phons[2]. As an example, a tone burst 100 msec long will sound about 12 dB louder than the same tone burst with a duration of 2 msec.

Very short transient components in a musical program can be considerably louder than the steady-state program level without sounding louder. One of the reasons the standard VU meter is preferred for monitoring signal levels is the fact that its ballistics conform fairly closely to the integrating effect of the ear. Figure 3-5 shows the response of the VU meter as a function of signal duration.

As we shall see in later chapters, peak reading meters are essential in maintaining overall integrity in a signal channel, but there is no question that the venerable VU meter corresponds more closely in its indications to how we hear.

MASKING

Masking phenomena occur in many ways. We know that loud sounds tend to mask softer ones. Higher frequency sounds are easily masked by lower frequency ones; but the reverse is not true. Relatively high levels of distortion at high frequencies are easily masked by the low frequency content of wide-band music. On the other hand, an unaccompanied chorus must be reproduced or reinforced with very low distortion, since there is little in the way of low frequency program to mask intermodulation distortion products.

CRITICAL BANDWIDTH

While the ear can detect relatively small differences in the pitch of individual pure tones, it has difficulty in separating two closely spaced pure tones. Assume that we have two sine wave oscillators and that they are both tuned initially to the same frequency. One of these oscillators will remain fixed, and the other will be variable. When we alter the frequency slightly, the listener will still hear only one tone. It will have a beat to it equal to the difference between the two frequencies. As we separate the tones further, the listener still detects only one tone, and the frequency of this single tone will be approximately half-way between the two frequencies being sounded. The listener will continue to hear a beat determined by the difference between the two frequencies.

With further separation of the frequencies up to about 10 Hz, the beat will turn into a certain roughness in the sound, but there will still be only one tone perceived by the listener. Finally, when the two tones are far enough apart, the listener will begin to hear them separately. The range over which the fusion of the two tones into a single tone is perceived is known as the critical band. Figure 3-6, after Roederer[3], shows the process we have described.

Another definition of the critical band is shown in Figure 3-7. Here, the critical bandwidth is defined as the minimum bandwidth of a noise signal which just masks a sine wave at its center when both the noise and the sine wave have the same power.

Both definitions give values of critical bandwidth which are quite close, and over the frequency range up to about 1 kHz, critical bandwidth is approximately equal to one-third octave. Above 1 kHz, critical bandwidth narrows to about one-quarter octave.

The importance of critical bandwidth lies in the fact that it is the maximum bandwidth over which the ear does not normally separate frequency components. Loudness within a critical band, for example, does not depend upon the levels of individual components, but rather on the RMS value of all components in the band. Variations in frequency response within a critical band are not normally heard as such, and variations in frequency response, if they are confined to critical bands,

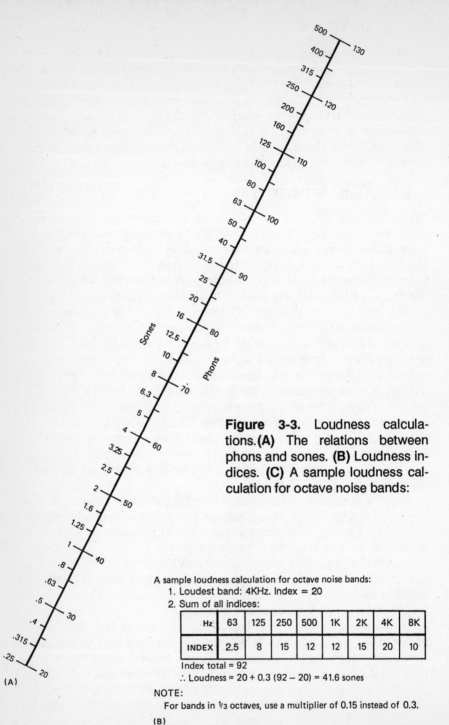

Figure 3-3. Loudness calculations. **(A)** The relations between phons and sones. **(B)** Loudness indices. **(C)** A sample loudness calculation for octave noise bands:

A sample loudness calculation for octave noise bands:
1. Loudest band: 4KHz. Index = 20
2. Sum of all indices:

Hz	63	125	250	500	1K	2K	4K	8K
INDEX	2.5	8	15	12	12	15	20	10

Index total = 92
∴ Loudness = 20 + 0.3 (92 − 20) = 41.6 sones

NOTE:
For bands in ⅓ octaves, use a multiplier of 0.15 instead of 0.3.

(B)

(A)

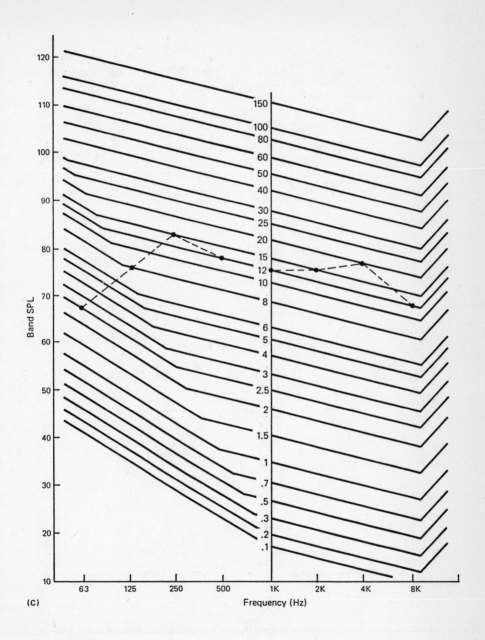

(C)

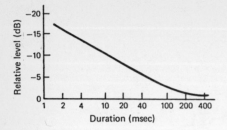

Figure 3-4. The dependence of loudness on tone duration.

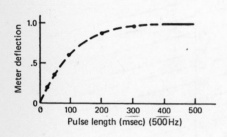

Figure 3-5. VU-meter ballistics.

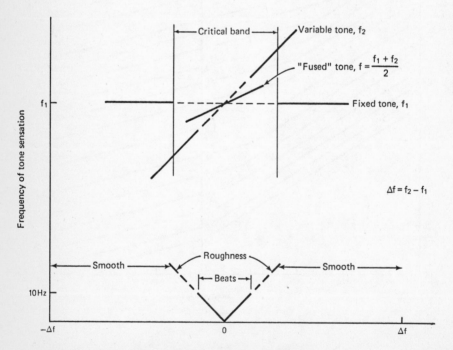

Figure 3-6. Critical bandwidth (data after Roederer).

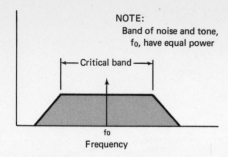

NOTE:
Band of noise and tone,
f₀, have equal power

Figure 3-7. Critical bandwidth.

will not usually be objectionable. The obvious exception here would be an extremely narrow, high peak in response, which would be obvious to the listener because of its effect on overall sound coloration.

PITCH PHENOMENA

For the most part, pitch is a function of frequency; the higher the frequency, the higher the pitch. However, under test conditions, most listeners will perceive subjective pitch, measured in *mels*, as shown in Figure 3-8. While a musician will relate two-to-one frequency ratios to judgements of twice the pitch, listening with pure tones will lead to other results, as the figure shows.

There is also a dependence of pitch on level. In general, frequencies below 1 kHz will appear to become lower in pitch with an increase in level, while above 1 kHz increases in level result in an increase in pitch. This data is shown in Figure 3-9.

TEMPORAL MASKING

The masking that we will be most concerned with in sound reinforcement is that due to time delay, or temporal displacement of sound stimuli. The phenomenon that we will be dealing with is known as the

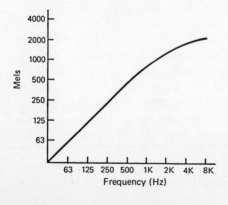

Figure 3-8. Subjective pitch (mels) vs. frequency.

59

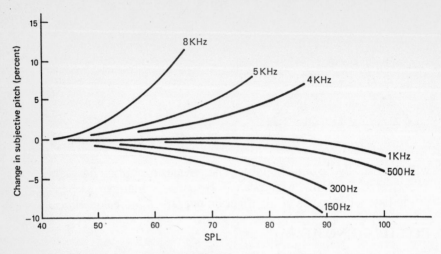

Figure 3-9. Pitch as a function of level.

precedence effect. Haas has described it in detail[4], and the phenomenon is often called the Haas effect.

While it may be shown in many forms, the graph shown in Figure 3-10A gives a good picture of the precedence effect. Consider two sound sources, both presenting the same program. When there is no delay between them, the listener will hear them equally, assuming that they are operating at the same level.

Now, let us put a variable time delay in line with one of the sources. As we gradually increase the amount of delay, the listener begins to localize the sound as originating at the undelayed source. For a given amount of delay in the zero-to-five msec range, the level of the delayed source can be increased up to 10 dB, relative to the undelayed signal, and balance will be subjectively restored. As the delay is increased beyond 5 msec, a constant 10-dB difference will restore the balance. As the delay increases beyond about 25 or 30 msec, the listener will begin to hear two distinct sounds, and there will be considerable difficulty in understanding speech.

The experiments shown in Figures 3-10B through E show these phenomena as they are normally employed in sound recording. In sound reinforcement, one of the most useful applications of the precedence effect is in such a system as shown in Figure 3-11. Here, a listener is seated some distance from a talker. Without reinforcement, the listener may have a difficult time understanding what is being said, unless the acoustics are adequate and the talker is speaking in an elevated voice. If reinforced sound is delayed suitably, then a small loudspeaker placed close to the listener will not be heard as such, but it will provide the necessary loudness to improve intelligibility. The

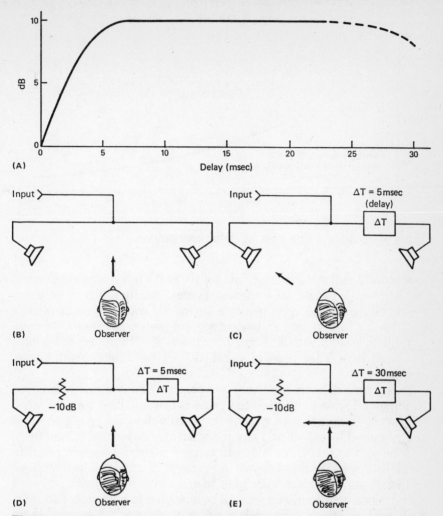

Figure 3-10. The precedence effect. **(A)** Delay versus amplitude imbalance relationships for precedence effect. **(B)** Phantom image in front. **(C)** Phantom image to left. **(D)** Delay plus attenuation restore the image to the center. **(E)** Increased delay spreads the image.

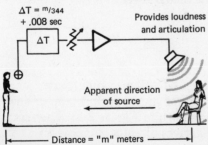

Figure 3-11. The precedence effect in sound reinforcement.

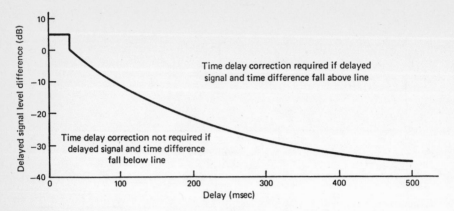

Figure 3-12. Doak & Bolt data for delay interference.

listener will still localize the sound source at the talker's position, while the delayed loudspeaker improves overall intelligibility. The delay should be calculated to compensate for the natural delay path, plus an additional 8 to 16 msec to allow the level of the loudspeaker to be comfortably increased without being heard as such. The approach we have described here is the basis of so-called "pew-back" systems in houses of worship.

Today, time delay is relatively inexpensive, and it is routinely specified in sound systems which, a few years ago, would have been designed without it. In many large systems, especially those set up out of doors, delays caused by echoes and auxilliary arrays are inevitable, and there may be no way that time delay can be used to correct a given problem for all listeners. The chart shown in Figure 3-12 was developed by Doak and Bolt[5], and it enables a designer to determine when a given delayed signal, however it comes about, will be annoying to a listener. The chart deals with relatively large delays up to as great as 0.5 second. As an example of how the chart is used, note that for a delay difference of 200 msec, the delayed source will have to be about 22 dB lower in level if it is not to be annoying to more than 10% of the listeners. A signal delayed 500 msec will have to be some 35 dB lower in level if it is not to annoy more than 10% of the listeners.

SUBJECTIVE ASPECTS OF PERFORMANCE SPACES

It is important that the student of sound reinforcement have some knowledge of what makes a room good for music performance. What the listener hears in a concert hall is shown in Figure 3-13. There is a direct sound component, that which arrives directly from the event on

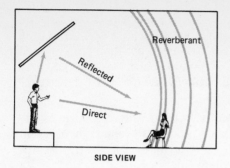

SIDE VIEW

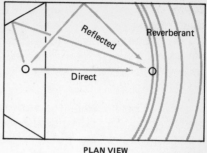

PLAN VIEW

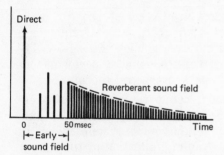

stage. It is followed by numerous discrete reflections, mainly from the side walls. Finally, after some time interval, there is the fairly diffuse reverberant field, which we studied in Chapter 2.

The time interval between the arrival of the direct sound and the onset of the reverberant field is known as the early sound field, and it is the precise structure of the early sound field which gives the listener a subjective feeling for the actual size of the room. In most concert halls, the early sound field lasts approximately 50 msec, but we note that there is no precise boundary between it and the onset of the reverberant field. Performers also benefit from initial reflections, inasmuch as they provide essential cues for good ensemble.

In a good concert hall, the initial time gap (the time between the perceived direct sound and the first reflection) should be no more than 25 msec.

The aim of good acoustical design is to balance the pattern of early reflections, and the directions they arrive from, with the development of the reverberant field in the room. In general a room is designed for a specific musical purpose, and attempts at multi-purpose designs have not, for the most part, been successful. There is no one way to design a performance space, and there are often many esthetic tastes to be satisfied. There is a rich vocabulary which the student should be familiar with as he learns more about this subject, and we can only cover a few elements of it here.

Intimacy: A room is intimate if the listener feels close to the players. Such rooms are small, seating perhaps 1200 at the most. The initial time gap may be as short as 15 or 20 msec, and reverberation time will be between 1.0 and 1.5 seconds in the midband. Careful control of early reflections maintains the feeling of intimacy, even toward the rear of the house.

Liveness: This is associated with relatively large amounts of reverberant energy in the midband (although not necessarily long reverberation time).The wood and plaster features characteristic of eighteenth century structures contribute heavily to this.

Warmth: This is a characteristic of most large, successful concert halls, and it implies a good amount of reverberant energy at low frequencies, with reverberation time at low and mid-frequencies in the 1.5-to 1.8-second range, such rooms lend themselves to convincing performances of the bulk of nineteenth century romantic orchestral literature.

Since sound reinforcement in performance spaces is merely an extension of the performers themselves, the attributes of a hall, good or bad, will usually impose themselves on the reinforcement system as well. In recent years, there has been considerable application of time delay and artificial reverberation, both of extremely high quality, in correcting some of the shortcomings of less-than-perfect performance spaces. These techniques are often incorporated into the main reinforcement system so successfully that the notion of a first-class multi-purpose hall has more credibility now than it has ever had. We will discuss these applications in detail in a later chapter.

CHAPTER 3:
References:
1. D. Robinson and R. Dadson, *British Journal of Applied Physics*, Vol. 7, p. 166 (1956).
2. C. Harris, *Handbook of Noise Control*, p. 8-13, McGraw-Hill, New York (1979).
3. J. Roederer, *Introduction to the Physics and Psychophysics of Music*, pp. 24-32, Springer-Verlag, New York (1973).
4. H. Haas, "The Influence of a Single Echo on the Audibility of Speech," *J. Audio Eng. Soc.*, Vol. 20, pp. 145-159 (1972).
5. R. Bolt and P. Doak, "A Tentative Criterion for the Short Term Transient Response of Auditoriums," *J. Acoustical Soc. Am.*, Vol. 22, pp. 507-509 (1950).

Recommended Reading:
1. L. Beranek, *Music, Acoustics, and Architecture*, John Wiley & Sons, New York (1962).
2. L. Doelle, *Environmental Acoustics*, McGraw-Hill, New York (1972).
3. H. Kuttruff, *Room Acoustics*, Applied Science Publishers, London (1979).
4. F. Winckel, *Music, Sound, and Sensation*, Dover Publications, New York (1976).
5. Various, *Halls for Music Performance*, Acoustical Society of America, New York (1982).

High-Frequency Systems

INTRODUCTION

Most large-scale sound reinforcement systems are at least of 2-way design, and the high-frequency (HF) portions of these systems usually make use of horn-compression driver combinations. More rarely, we will see HF systems made up of direct radiator, or cone, elements. The frequency range HF devices cover is nominally from 0.5 to 1 kHz at the lower bound, and up to 10 to 15 kHz at the upper bound. In some special cases requiring increased output above 8 kHz, an assemblage of small compression drivers known as ring radiators may be used to increase output capability.

We will discuss the various methods for power rating of these elements, as well as ways of protecting them from electrical abuse in normal usage. Finally, there will be considerable discussion of how HF directional patterns are modified when multiple radiators are closely spaced.

COMPRESSION DRIVERS

High-frequency horns behave in part as acoustical transformers, converting high-pressure sound at low volume-velocity at the throat into low-pressure sound at high volume-velocity at the mouth of the horn. This relationship is shown in Figure 4-1.

An analogy may be made between acoustical power and electrical power, with sound pressure (normally measured in pascals) being the analog of potential and volume-velocity (normally measured in m³/sec) the analog of current. Just as an electrical transformer maintains equal products of current and potential at its input and output, the horn

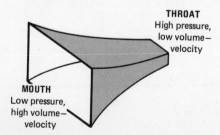

THROAT
High pressure,
low volume—
velocity

MOUTH
Low pressure,
high volume—
velocity

Figure 4-1. Pressure and volume velocity relationships in a horn.

maintains equal products of pressure and volume-velocity at its throat and at its mouth.

The term volume-velocity may be new to many readers; it has reference to the bulk displacement of air, not the speed of sound.

The horn should be driven by a transducer capable of producing high pressures at the throat. Such transducers are known as compression drivers, and typical models can exhibit electro-acoustical conversion efficiencies on the order of 20 to 30% when coupled to an appropriate acoustical load.

CONSTRUCTION DETAILS

Figure 4-2 shows details of compression drivers. The driver shown at A is typical of the basic design of compression drivers as developed by Western Electric some 50 years ago. Typical diaphragm diameters vary from 44.5 mm (1.75 in.) up to about 100 mm (4 in.), and the materials used for diaphragm construction have included aluminum, beryllium, titanium, phenolic-impregnated linen, and various combinations of metals with plastic surround portions.

The voice coil is usually made of aluminum ribbon wire, edge-wound for higher packing density, and it is placed in a magnetic gap with flux densities in the range of 1.7-to-2 tesla (17,000-to-20,000 gauss). Magnet materials include Alnico V, an alloy of aluminum, nickel, and cobalt, as well as various ferrite ceramic materials.

The diaphragm normally fits within 1 mm (.04 in.) of the phasing plug. The phasing plug has a set of slits, whose area at the diaphragm surface is about one-tenth the area of the diaphragm itself. The ratio of these areas, known as the loading factor of the driver, affects the efficiency of the device.

The design shown at B has radial slits in the phasing plug, while the more common form, shown at A, has annular, or circular, slits.

RADIATION RESISTANCE

Figure 4-3A shows an equivalent circuit which describes the mid-band efficiency of the driver. R_E represents the resistance of the voice coil in ohms. The radiation resistance, R_{ET}, of the driver coupled to a horn or other load, represents the electrical resistive component seen at the driver terminals that produces acoustic power output. It is given by the following equation:

$$R_{ET} = S_T (Bl)^2/\rho_0 c\ S_D^2, \tag{4-1}$$

where S_T = slit area in m²,
$\quad$ B = magnetic flux density in teslas,
$\quad$ l = conductor length in meters
$\quad \rho_0 c$ = characteristic acoustical impedance of air,
$\qquad$ equal to 415 N·sec/m³.

66

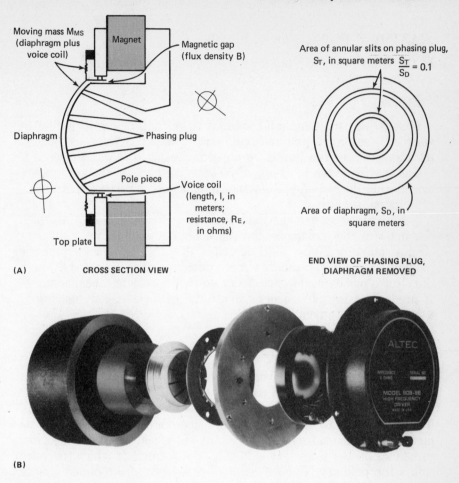

(A) CROSS SECTION VIEW

END VIEW OF PHASING PLUG, DIAPHRAGM REMOVED

(B)

Figure 4-2. Construction details of compression drivers.

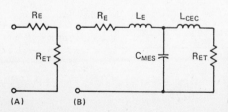

Figure 4-3. Equivalent circuits for **(A)** mid-frequency and **(B)** high-performance of a compression driver.

67

MAXIMUM POWER TRANSFER

In the following equation, power transfer is maximized at 50% when $R_E = R_{ET}$:

$$\text{Efficiency (\%)} = [2\, R_E R_{ET}/(R_E + R_{ET})^2] \times 100 \qquad (4\text{-}2)$$

Under this condition, half the input power will be dissipated in the voice coil resistance, and half will be radiated as useful sound output. (Occasionally, in the literature on compression drivers, reference is made to efficiencies in excess of 50%. In these cases, the authors are considering the radiation resistance term alone, in relation to its maximum possible value.)

The frequency response will be uniform up to the frequency at which the mass of the moving system becomes dominant. The equivalent circuit shown in Figure 4-3B describes this. C_{MES} is dependent on the moving mass and represents a mechanical quantity reflected through the electro-mechanical system as an equivalent electrical capacitance. It produces a high-frequency roll-off of 6 dB/octave, commencing at a frequency, f_{HM}, given by:

$$f_{hM} = (Bl)^2/\pi R_E M_{MS} \qquad (4\text{-}3)$$

where M_{MS} is the mass of the moving system in kilograms.

In most compression drivers, f_{HM} is in the region of 3 to 3.5 kHz. Two other elements in the equivalent circuit shown in Figure 4-3B may affect HF response. L_E represents the inductance of the voice coil; it can effectively be nulled by plating a silver or copper ring on the pole piece. This "shorted turn" in the vicinity of the voice coil acts like a low resistance on the secondary side of a transformer, of which the voice coil is the primary. The low value of resistance of the plated ring is reflected through, and swamps out the reactance of the inductive element.

The L_{CEC} element depends on the volume of the air cavity between the diaphragm and phasing plug. Keeping the diaphragm-phasing plug spacing as small as possible, consistent with excursion requirements at high power input levels near the lower frequency cut-off of the system, will minimize its effect in the audible HF range.

In some compression driver designs, the range above 10 kHz can be

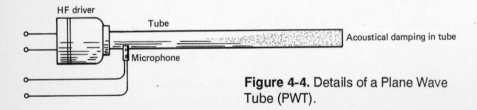

Figure 4-4. Details of a Plane Wave Tube (PWT).

enhanced through the action of secondary resonances in the diaphragm and compliance. There are no simple equivalent circuits which describe these effects. As we shall shortly see, these secondary resonances can be manipulated to yield overall improvements in response with negligible performance trade-offs.

MEASUREMENT OF COMPRESSION DRIVERS

In order to minimize the aberrations and directional effects of horns, compression drivers are usually measured on a plane wave tube (PWT). Figure 4-4 shows a typical PWT. The driver is attached at one end, and a test microphone is located close by. The rest of the PWT is filled with a tapered wedge of fiber glass, or some other rugged and absorptive material, which attenuates the sound sufficiently so that standing waves, or reflections, in the PWT are minimized.

The equation which relates the measured sound pressure in the PWT and the acoustical intensity (power per unit area) in the PWT is given:

$$P_{PWT} = \sqrt{\rho_o c P_A / S} \qquad (4\text{-}4)$$

where P_A is the acoustic power in watts, and S is the cross-sectional area of the PWT in M^2.

Solving this equation and converting to level in dB, we note that one acoustical watt in a 25.4 mm (1 in.) PWT will produce a pressure of 153 dB-SPL. As stated earlier, the maximum possible efficiency of a compression driver is 50%. Therefore, we should not expect to see one-watt response on a PWT of this diameter exceed 150 dB-SPL.

Figure 4-5 shows the response of a typical compression driver with a 44.5 mm (1.75 in.) diaphragm. Here, the electrical power input is 1 milliwatt, 30 dB lower in level than one watt. Dashed lines show the transition between the mid-frequency (MF) region and the HF region, determined by the moving mass. The region affected by secondary resonances is also indicated. Note that the mid-frequency SPL is 117 dB, three dB below the possible maximum of 120, (150-30), dB. The indicated efficiency here would be one-half of 50%, or 25%. Note that the compression driver with substantial mid-band efficiency can have an efficiency of only a few percent in the 10 kHz range.

POWER RATINGS OF COMPRESSION DRIVERS

At high frequencies, where diaphragm excursion is slight, the power rating of a driver will be determined by its ability to dissipate the heat generated in the voice coil. This is the so-called thermal power rating of the device. At frequencies in the 500-to-800 Hz range, the displace-

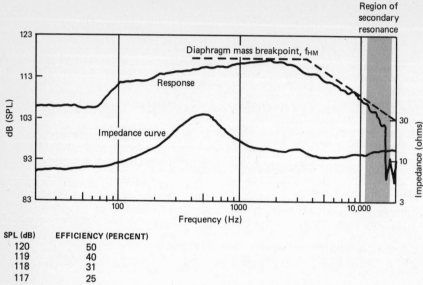

SPL (dB)	EFFICIENCY (PERCENT)
120	50
119	40
118	31
117	25
116	20
115	16
114	12.5
113	10
112	8
111	6.3
110	5
109	4
108	3.1
107	2.5

Figure 4-5. Response of a compression driver (1.75-inch diaphragm diameter), 1 milliwatt input, 2.54 cm plane wave tube.

ment of the diaphragm will limit the input power, and catastrophic failure of the diaphragm will occur if it strikes the phasing plug. In addition, multiple flexures of the diaphragm may result in cumulative metal fatigue such that after years, or perhaps only months of operation (depending on input level), a diaphragm will simply fracture in the surround region, even though specific displacement or thermal limits may not have been exceeded.

Aluminum is quite vulnerable to fatigue, while materials such as beryllium, titanium, and phenolic-impregnated linen are much less so.

Most manufacturers carefully state their recommended input power ratings based on the recommended lower cross-over frequencies and the slopes of the cross-over transitions. A few manufacturers give additional data in the form of a *Safe Operating Area* (SOA) graph, as shown in Figure 4-6. The SOA graph indicates that a driver may easily handle short bursts of power which, on a steady-state basis, would burn out the voice coil. The graph takes into account the thermal inertia of the voice coil in the form of an allowable duty cycle at elevated power inputs. The assumption is made that, in the shaded region of the graph, displacement limits will not be exceeded.

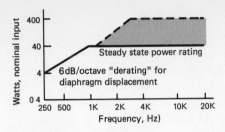

Figure 4-6. A Safe Operating Area (SOA) graph for a compression driver. In the cross-hatched region the driver may sustain momentary power bursts up to ten times the steady-state rating, providing a duty cycle of 1-to-10 is not exceeded.

COMPARING MANUFACTURER'S SENSITIVITY DATA

Most high-quality professional compression drivers have sensitivities which are within a dB or so of each other. A general standard in the industry is to present PWT data measured on a 25.4mm (1 in.) tube with a nominal power input of 1 milliwatt ($E_{in} = \sqrt{(.001)/Z_{nom}}$).

Some manufacturers use tubes of different diameter, necessitating conversions. For example, one manufacturer specifies a PWT with a cross-section of 19mm (0.75 in.). The reduced area increases the level by 2.5 dB, and this must be taken into account in making sensitivity comparisons.

For large drivers, 50mm (2 in.) PWTs are often used, since measuring them on a 25.4mm PWT would require a necking down of the tube prior to the point where the measurement is actually made. As a rule, manufacturers of these larger devices will make the conversion for the user, referring the levels to their equivalents on a 25.4 mm tube.

Some manufacturers prefer to present sensitivity data with the driver mounted on a given horn. The standard here is the SPL referred to a distance of one meter with power input of one watt. If such a specification does not precisely indicate the directivity index of the horn as a function of frequency, then the rating is practically useless in making competitive comparisons. The following equation can be used to determine the efficiency of a driver specified in this way:

$$10 \log (\text{efficiency}) = \text{Sensitivity (1 watt, 1 meter)} - 109 - DI \qquad (4\text{-}5)$$

As an example, let us assume that a particular horn-driver combination is rated at 113 dB SPL, one watt at one meter, and that the DI at some particular frequency is 11 dB. Then:

$$10 \log (\text{efficiency}) = 113{-}109{-}11$$
$$10 \log (\text{efficiency}) = {-}7 \text{ dB}$$

Efficiency $= 10^{\frac{-7}{10}} = 0.2$

Thus, the efficiency of this horn-driver combination at the particular measurement frequency is 20%.

SENSITIVITY MEASUREMENTS

While most manufacturers specify the sensitivity of horn-driver combinations as the output level for one watt input at a distance of one meter, the actual measurements on the device are not made so close to the mouth of the horn. The usual procedure is to make the sensitivity measurement at a distance of, say 10 meters, and then add 20 dB to it to convert it, via inverse square law, to an equivalent 1-meter rating. The reason for this is simply that it is difficult to identify with precision the actual acoustic center of a horn-driver combination, and a small error in a close-in measurement distance could result in significant level errors.

Some manufacturers give sensitivity ratings at four feet rather than one meter. The conversion here is 1.7 dB; that is, a horn-driver combination with a sensitivity of 111 dB SPL, one watt at four feet, is equivalent to 112.7 dB, one watt at one meter.

THE ROLE OF SECONDARY RESONANCES

Figure 4-7 shows the response of three compression drivers mounted on the same horn. All three have a diaphragm diameter of about 100 mm (4 in.), but differ in diaphragm material and surround treatment. The JBL 2440 has a "half roll" surround which produces a gradual rise in response to about 9 kHz, above which point the response drops rapidly. The model 2441, through the use of a different surround treatment, distributes the secondary resonances in such a way that

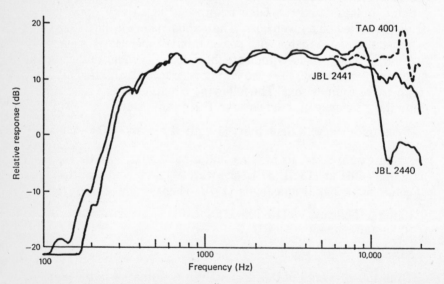

Figure 4-7. The role of secondary resonances.

smoother, but somewhat reduced, response is extended out to about 18 kHz. The TAD model 4001 driver has the same surround detail as the 2440, but it is made of a much stiffer material, beryllium. Because of the added stiffness, a response peak similar to that of the 2440 is observed, but it is raised about one octave to about 17 kHz.

Secondary resonances can be used in two ways: to elevate the response out to some cut-off frequency, or to extend the response beyond that cut-off frequency, but with some reduction in output.

HIGH-FREQUENCY HORNS

There are five basic types of horns we will discuss: multi-cellular horns, radial (sectoral) horns, horn-lens combinations, diffraction horns, and constant coverage horns. Together, they form the basis for most high-level sound reinforcement applications. While the newer constant coverage horns have solved most of the pattern control problems inherent in the older designs, there are certain applications in which the less-than-perfect characteristics of some of the earlier horns may be desired.

Ideally, a horn should efficiently couple the HF driver's output and produce smooth, extended response. It should also exhibit smooth dispersion over its operating range. Horns have become better in recent years, but there are still compromises in all designs.

EXPONENTIAL HORN THEORY

We will begin our survey of horns with a look at an infinitely-long exponential horn, as shown in Figure 4-8. The horn's cutoff frequency, f_c, should be at least one octave below the lowest frequency at which the horn is expected to operate.

The graph of Figure 4-8B shows the resistive and reactive loading of an infinitely-long horn. Note that when the ratio of f to f_c is 2 or greater, the leading effect is essentially resistive.

Of course, there are no infinite horns, and the typical horn will present X_t and R_t curves that will have "lumps" in them, due to reflections from the discontinuity of the mouth back to the driver. These conditions are shown at C and D in the figure.

To determine the shape of the horn, we begin by calculating the flare constant, m, which determines how rapidly the horn flares out as the distance from the throat increases. The equation is:

$$m = 4\pi f_c/c \qquad (4\text{-}6)$$

where m = flare constant, in meters^{-1},

f_c = cutoff frequency, and

c = velocity of sound in meter/sec.

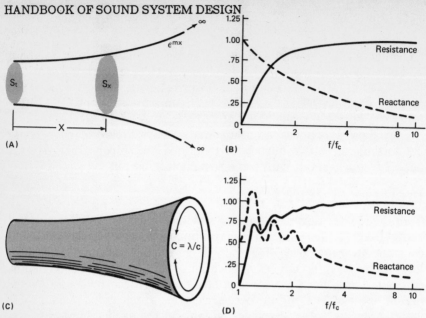

Figure 4-8. The exponential horn. **(A)** An infinitely long exponential horn. **(B)** Radiation resistance and reactance for and infinite exponential horn. **(C)** A finite horn. **(D)** Radiation resistance and reactance for a finite horn.

At any point along the length of the horn, its cross-sectional area may be found from the equation:

$$S_x = S_t e^{mx} \qquad (4\text{-}7)$$

where

S_x = the cross-sectional area at a distance, x,
S_t = the cross-sectional area at the throat,
e = 2.718
m = flare constant, and
x = distance from the throat in meters.

If the circumference, c_m, of the horn mouth, divided by the longest wavelength to be reproduced, is greater than about 3 (i.e., $c_m/\lambda_{max} > 3$), then the horn will behave approximately as though it were infinite, with a quite smooth radiation resistance.

MULTI-CELLULAR HORNS

Although a simple exponential horn provides ideal loading for a driver, it does tend to focus, or beam, high frequencies along the axis of the horn. The multi-cellular horn, developed in the early thirties by Wente and Thuras[1], was the first attempt to overcome the problem of HF beaming. A group of exponential horns, or "cells," were clustered together, as shown in Figure 4-9A, and each cell was expected to control radiation in its own direction. The ideal performance is shown at B.

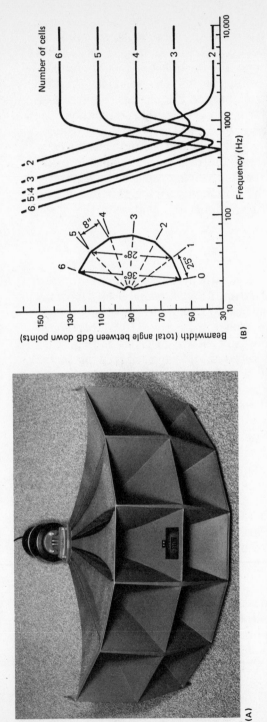

Figure 4-9. The multicellular horn. **(A)** A typical 2 × 5 multi-cellular horn (Altec photo). **(B)** Theoretical beamwidth for multi-cellular horns (data from Beranek, *Acoustics*, McGraw-Hill, a New York, 1954). **(C)** Polar response of a 2 × 5 multi-Cellular horn at 2 kHz (horizonal, dashed line; vertical, solid line). **(D)** Polar response of a 2 × 5 multi-Cellular horn at 10 kHz (horizonal, dashed line; vertical, solid line).

Note that there is narrowing of the pattern, for all configurations, in the 500 to 800 Hz range. This so-called mid-range narrowing is the result of the group of cells behaving as a single, large mouth at the frequency whose wavelength is equal to the effective mouth width.

Another problem with multi-cellular horns is the tendency for them to "finger" along the individual cells at high frequencies. Figures 4-9C and D show this quite clearly for a two-by-five cell combination. The depth of the fingering approaches 10 dB at the highest frequencies, and the fingering is most pronounced along the septum between cells.

Multi-cellular horns have served the professional sound industry well for many years, but they are rapidly being replaced by the newer constant coverage horns. Even the motion picture sound industry, long the hold-out for the multi-cells, is making the change to the newer designs.

RADIAL HORNS

These devices are so-named because, when viewed from above, they resemble a sector of a circle with straight radial boundaries. They may also be referred to as sectoral horns. Top and side views of a typical 90-by-40 degree radial horn are shown in Figure 4-10A. The typical directional properties of such a horn are shown at B. Note, as in the case of the multi-cellular horn, that there is mid-range narrowing in the horizontal direction at the frequency whose wavelength is equal to the mouth width. Further, because the vertical cross-section of the radial horn is exponential, the vertical response narrows progressively with frequency over the entire range.

For many applications, the vertical pattern narrowing of radial horns is an advantage. Where horizontal control is the main concern, the rising DI of the horn acts to "equalize" its response by an amount equal to the DI curve itself. The typical DI for a radial horn is shown in Figure 4-10C.

Figure 4-11 shows typical radial designs in use today. The most common forms of these devices are those that offer nominal 90-by-40 degrees, 60-by-40 degrees, and 40-by-20 degrees.

ACOUSTIC LENSES

Like the multi-cellular horn, the acoustic lens represents an attempt to overcome the tendency of exponential horns to beam at high frequencies. The acoustic lens was described by Kock and Harvey of Bell Telephone Laboratories in the forties[2], and later studies were undertaken by Frane and Locanthi[3]. There are two basic types of lenses—slant plate and perforated plate, as shown in Figure 4-12. By either action, there is a shorter path length through the center of the lens than at its edges. Thus, sound

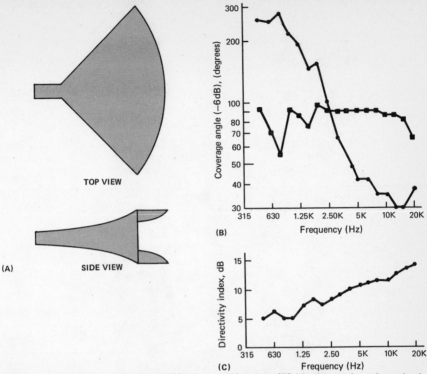

Figure 4-10. The radial horn. **(A)** Views of horns. **(B)** Horizontal and vertical beamwidth of a typical 90 x 40 radial horn (JBL data). **(C)** Directivity index (D1) for the radial horn whose beamwidth data is shown in **Figure 3(B)**.

waves exiting at the center are ahead of those exiting at the sides and tend to fan out, offering wider coverage.

The perforated plate lens is normally circular, producing a symmetrical pattern about the horn's axis. The slant plate lens will always have distinctly different horizontal and vertical patterns. Figure 4-13 shows beamwidth data on a typical slant plate lens. Because there is no lens action in the vertical plane, the coverage in that plane shows the typical beaming at high frequencies of an exponential horn. In the horizontal plane, the coverage is nearly ideal.

Lenses are manufactured by only a few companies, and they are used most often in music reinforcement or monitoring systems. Since lenses are used as diverging elements, it is not uncommon to see them employed to provide quite wide coverage (up to 120 degrees in the horizontal plane). Thus, they are best specified for fairly short-throw applications. It is the observation of many users that acoustic lenses are "soft-edged"; that is, they do not fall off as rapidly beyond their nominal coverage angles as do other types of horns. Again, this may favor them for musical applications as opposed to speech applications.

Figure 4-14 shows a group of typical commercial acoustical lenses.

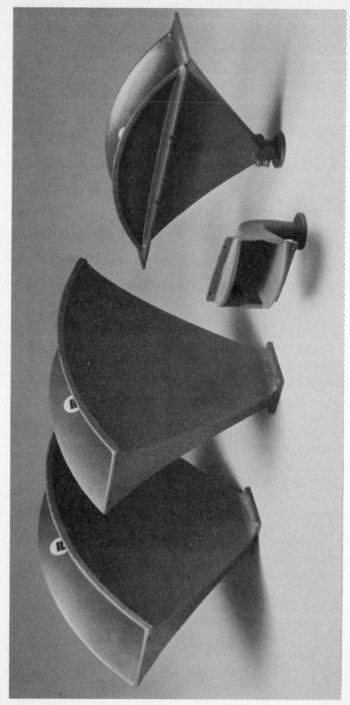

Figure 4-11. Typical commercial radial horns. (JBL Photo).

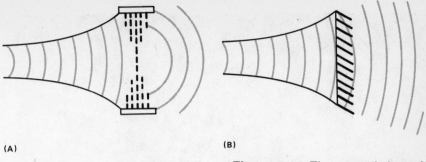

(A)

(B)

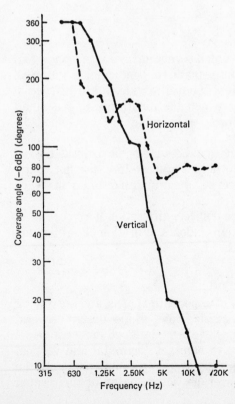

(C)

Figure 4-12. The acoustic lens. **(A)** A perforated plate acoustic lens. **(B)** A slant plate acoustic lens, side view. **(C)** A slant plate acoustic lens, top view.

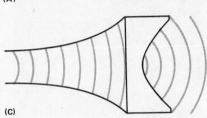

Figure 4-13. Beamwidth data for a slant plate acoustic lens.

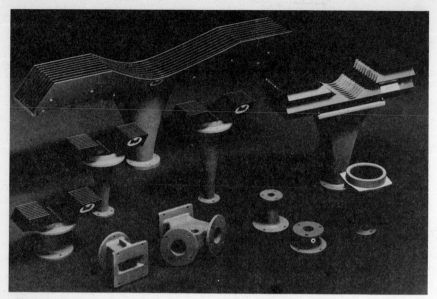

Figure 4-14. Commercial acoustic lenses. (JBL Photo).

DIFFRACTION HORNS

The diffraction horn has a mouth which is very narrow in one plane and usually fairly wide in the other. In the plane perpendicular to the small mouth opening, the coverage angle will be quite wide, due to diffraction effects, for those frequencies for which the mouth opening is small compared to the wavelength. At progressively higher frequencies, the coverage will become narrower.

In the other plane, the coverage may be wide or narrow, as the designer wishes. In the model shown in Figure 4-15A, the horizontal coverage angle is just over 130 degrees. Beamwidth data for this horn is shown in Figure 4-15B.

As with all other wide-coverage devices, diffraction horns are most useful for music reproduction systems indoors, or in other close quarters where narrow coverage is not desired.

CONSTANT COVERAGE HORNS

Departing from the simple theory of exponential horns, and borrowing from the practice of electromagnetic wave guide design, several families of constant coverage horns have been designed in recent years. While most conventional horns have nominal horizontal and vertical coverage angles which they may satisfy over only a small portion of the frequency range, the constant coverage horns maintain their rated

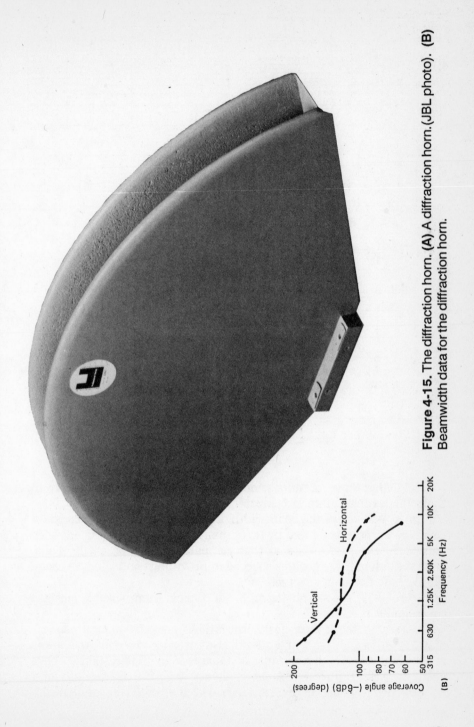

Figure 4-15. The diffraction horn. (A) A diffraction horn. (JBL photo). (B) Beamwidth data for the diffraction horn.

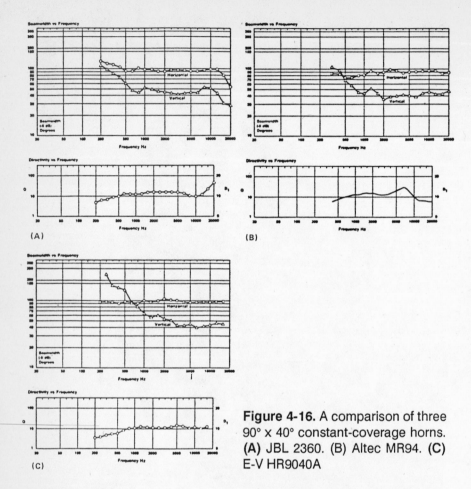

Figure 4-16. A comparison of three 90° x 40° constant-coverage horns. **(A)** JBL 2360. (B) Altec MR94. **(C)** E-V HR9040A

coverage angles over a quite wide band of frequencies, typically, from the 800 Hz range up to 12.5 kHz.

Figure 4-16 shows the beamwidth data for 90-by-40 degree constant coverage horns as produced by three manufacturers. The data has been plotted on the same scale so that comparisons can be easily made. Note that all of these designs succeed in providing controlled coverage over a wide frequency range.

Figure 4-17 shows photographs of typical commercial constant coverage horns.

The great virtue of these new horns is that smooth frequency response may be realized both on- and off-axis. But in order to do this, the power response of the HF driver must be equalized for flat response above about 3.5 kHz. Figure 4-18A shows the on- and off-axis response for a 90-by-40 degree constant coverage horn with the HF driver so equalized.

Figure 4-17. Commercial constant-coverage horns. (JBL photo).

Figure 4-18B shows what typically happens when a nominal 90-by-40 degree radial horn is equalized for flat on-axis response. Note that HF response falls off for even small off-axis horizontal angles.

The typical DI range for constant coverage horns in their usual configurations is:

90-by-40 degrees   10-12 dB
60-by-40 degrees   12-14 dB
40-by-20 degrees   15-17 dB

New designs of smaller horns incorporate many of the virtues of the larger constant coverage horns illustrated here. As a general rule, they all exhibit excellent horizontal control, but maintain their vertical pattern control only down to 1.5 or 2 kHz, due to their fairly narrow vertical mouth openings.

DISTORTION IN HORN SYSTEMS

With well-designed HF drivers in good operating condition, distortion in HF horn systems will result only from thermodynamic air overload rather than from non-linearities in the moving system. At normal operating levels, the percent of second harmonic distortion to be expected from a horn system may be found from the following equation:

$$HD_2 = 1.73 \, (f/f_c)\sqrt{I_T} \times (10^{-2}), \qquad (4\text{-}8)$$

83

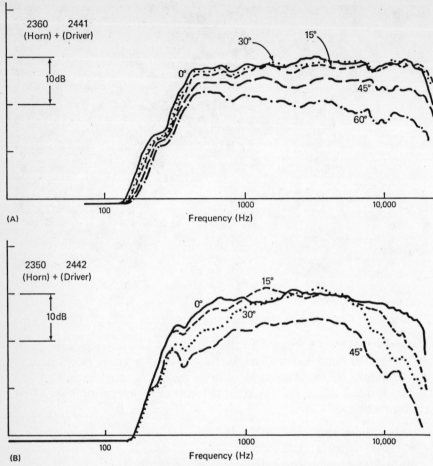

Figure 4-18. On- and off-axis response of constant-coverage and radial horns. **(A)** On-axis and off-axis response of a 90° by 40° constant coverage horn with equalized HF driver (data courtesy JBL and D.B. Keele). **(B)** On-axis and off-axis response of a 90° by 40° radial horn.

where HD$_2$ = percent second harmonic distortion,

      f = frequency of input signal,

      f$_c$ = horn's nominal cutoff frequency, and

      I$_T$ = intensity, in watts/m^2, at the diaphragm-phasing plug interface.

Note that the distortion is proportional to the driving frequency and inversely proportional to the cut-off frequency of the horn. Thus, horns with more rapid flare rates (higher cut-off frequencies) will generally produce less distortion than those designed to operate at lower frequencies.

84

Distortion is proportional to the square root of the intensity at the diaphragm-phasing plug interface, and this implies that the larger diameter diaphragm drivers will produce less distortion, when mounted on the same horn and driven to the same level, than will the smaller drivers. Further, each doubling of the output level will produce a 3-dB increase in distortion, relative to the fundamental; and raising the driving frequency one octave, keeping all else the same, will raise the second harmonic distortion by 6 dB relative to the fundamental.

Figure 4-19 shows the on-axis frequency response, with second harmonic distortion components, for a 60-by-90 degree constant coverage horn. At A, the horn is powered with one watt through the entire frequency range, and at B, the input power is 10 watts. A sample calculation of distortion at 1 kHz for a power input of 1 watt is given in the figure. Note that the agreement between the measurement and the calculated value is excellent. Distortion levels such as these are typical of most horn-compression driver systems.

As a practical matter, making distortion measurements in the field is a difficult job, due to ambient noise conditions. Normally, in a properly designed system, distortion will not be a problem, due largely to the fact that the frequency spectrum for speech, as well as most music, tends to be rolled off at high frequencies.

Distortion comparisons *must* be made at equal acoustical output levels—not at equal power input levels.

HIGH-FREQUENCY DRIVER PROTECTION

It is essential that all HF drivers be protected from inadvertent dc input through the use of a blocking capacitor. Where the driver is used with a high-level, passive dividing network, this protection is already present. But where the driver is a part of a bi-amplified system, the designer must put a capacitor in series with the driver at the HF amplifier's output. Figure 4-20A and B provide information on how to select the correct values. In general, a capacitor is selected whose reactance in ohms is equal to the nominal impedance of the driver at one-half the crossover frequency.

Not all system designers are in agreement on what kind of protection, if any, should be provided for actual program limiting because of over-driving of the system. If a system is adequately specified—and operated within the implied limits—then no program limiting should be necessary, other than routine monitoring of levels at the control console. But where inexperienced hands may be at the controls, some kind of limiting will be necessary. Figure 4-20C shows details of a circuit developed by Electro-Voice that rectifies high-signal currents, actuating a relay that attenuates the signal to the HF driver by some predetermined amount. This circuit provides for the setting of the threshold of action, and the action is frequency-selective so that poten-

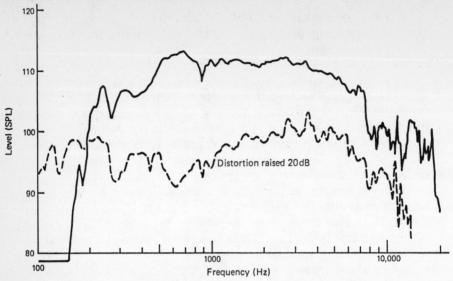

Distortion calculation:

1. JBL 2441: 1 Watt produces 148 dB-SPL in 25.4 mm PWT
2. Acoustical power is therefore 0.3 Watt
3. Area of diaphragm = $\pi (50)^2$ mm^2, or 7.6×10^{-3} m^2
4. Area of slits at phasing plug = $\frac{1}{10}$ x diaphragm area, or $.76 \times 10^{-3}$ m^2
5. $I_T = \frac{.3 \times 10^3}{.76} = 394$ Watts/m^2
6. $HD_2 = 1.73 \left(\frac{1000}{200}\right) \sqrt{394} \times 10^{-2} = 1.7\%$
7. f_c assumed to be approximately 200 Hz
8. Measured 2nd harmonic level is 37 dB below fundamental, or approximately 1.4%

(A)

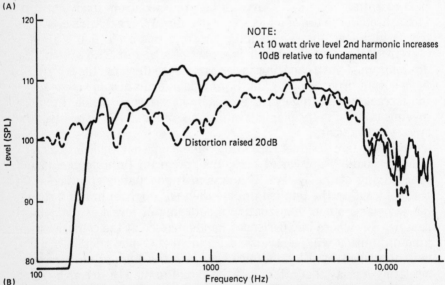

NOTE:
At 10 watt drive level 2nd harmonic increases
10 dB relative to fundamental

(B)

Figure 4-19. Distortion in compression driver systems. **(A)** A 60° by 40° constant coverage horn driven at 1 watt; the solid line is the fundamental, and the dashed line is the second harmonic (raised 20 dB). **(B)** A 60° by 40° constant coverage horn driven at 10 watt; note that at the 10-watt drive level the 2nd harmonic increases 10 dB relative to the fundamental.

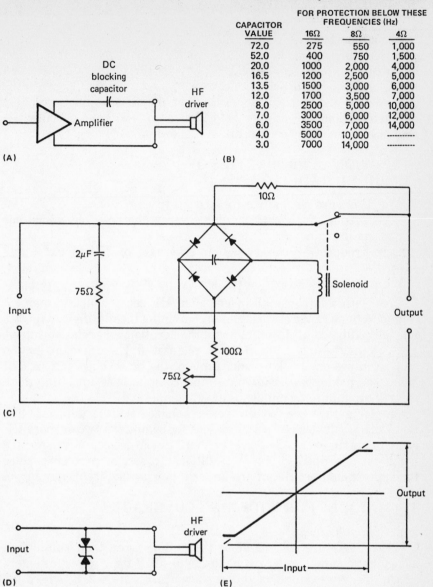

CAPACITOR VALUE	FOR PROTECTION BELOW THESE FREQUENCIES (Hz)		
	16Ω	8Ω	4Ω
72.0	275	550	1,000
52.0	400	750	1,500
20.0	1000	2,000	4,000
16.5	1200	2,500	5,000
13.5	1500	3,000	6,000
12.0	1700	3,500	7,000
8.0	2500	5,000	10,000
7.0	3000	6,000	12,000
6.0	3500	7,000	14,000
4.0	5000	10,000	----------
3.0	7000	14,000	----------

Figure 4-20. HF driver protection. **(A)** Use of a d.c. blocking capacitor. **(B)** Selection of the blocking capacitor, C: **(C)** A schematic of an Electro-Voice STR tweeter protector modified for use with compression drivers. **(D)** Zener diodes across the inputs of a compression driver.

tially damaging lower frequencies are sensed with a lower threshold than are high frequencies.

Finally, Figure 4-20D shows details of a Zener diode network placed across the terminals of a HF device. These diodes, once their conduction threshold has been exceeded, will provide a shunt path for the signal, effectively bypassing the driver. While this circuit produces considerable distortion while in operation, it may be argued that the distortion is preferable to a burnt-out voice coil. If used across a driver that is crossed over above, say, 7 or 8 kHz, then the distortion will probably not be objectionable—or even noticed as such. The input-output characteristic of the diode configuration is shown at E.

DIRECTIONAL PROPERTIES OF COMBINED RADIATORS

HF radiators are normally combined for three reasons:

1. To produce *increased* angular coverage beyond that which one device can provide.

2. To produce *decreased* angular coverage for "far-throw" applications.

3. To provide *increased* output level capability.

Even with a fairly small inventory of HF components, we can encounter a huge range of variables in combining them. Different models may be combined at different azimuth, elevation, and rotation angles, and they may be driven at different levels as well. The aim in this section will be to present to the reader some of the broad observations that have been made with relatively simple combinations, and from these we will attempt to formulate a few guidelines.

Simpler arrays are usually better behaved than larger ones, and there should always be logical reasons for combining two or more HF elements in an array. To a great extent, the larger constant coverage horns have simplified many traditional coverage problems, since their directional properties are uniform over a wide frequency range.

TECHNIQUES FOR WIDENING COVERAGES

1. *Splaying along the -6 dB axes:*

If two horns are splayed along their -6 dB axes, their summation along that common axis will be nearly 0 dB if their acoustic centers are in line. An example of this is shown in Figure 4-21B, where the horizontal coverage angle of splayed 90-degree horns has been extended to 180 degrees, as compared to the single horn shown at A.

Note that splaying has minimal effect on the vertical beamwidth of the array.

The stack-and-splay arrangement shown at C is better than that shown at B, since it produces less lobing in the horizontal polar patterns.

By extension, a pair of 60-by-40 degree horns would provide smooth 120-by-40 degree coverage if stacked and splayed horizontally by 60 degrees.

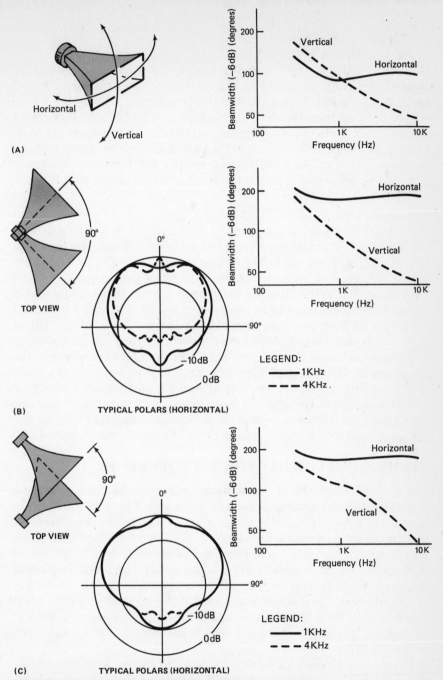

Figure 4-21. Techniques for widening coverage. **(A)** Beamwidth of a single horn. **(B)** Data for a splayed pair of 90° horns. **(C)** Data for a stacked and splayed pair of 90° horns.

2. *Cross-firing of horns:*

The smaller the combined horn mouth area of an array, the better behaved it will be. For increased vertical coverage, horns are often splayed vertically, as shown in Figure 4-22A. Note that the extended vertical beamwidth is nearly constant, broadening a bit in the 4 kHz range. At B, we show the same angular orientation, but with the drivers at the center. Note that the vertical pattern control is not as smooth as in the case shown at A. Lobing in the 1 and 2 kHz range is greater than the case shown at A. Finally, observe that the horizontal coverage provided by these two arrays is the same as with a single horn (Figure 4-21A).

3. *Requirements for near and far coverage:*

Figure 4-23 shows a side view of a pair of horns covering a long, narrow room. The far-throw horn has a 40-by-20 degree pattern, and the near-throw horn has a 90-by-40 degree pattern. The problem here is that the horns' outputs are not the same along their respective −6 dB axes, inasmuch as the DI of the long-throw horn is 18 dB, while that of the near-throw horn is 13 dB. In this case, we must consider the actual direct sound field levels as they exist in the seating area and adjust drive levels accordingly. As before, the designer has a choice of a cross-fired concave array or a convex array. In such an array as this, the conventional definitions of horizontal and vertical coverage angles do not apply, since the actual angular coverage is modified by the varying coverage distances required. In these cases, the best recommendations are to keep the splay angles as wide as possible, consistent with proper coverage, and to keep the combined mouth area as small as possible.

TECHNIQUES FOR NARROWING COVERAGE

In-line vertical stacks of horns have long been used for producing narrower vertical coverage than can be provided by a single horn. In the case of the older radial designs, which tended to lose vertical coverage due to their narrow vertical mouth dimension, the larger effective mouth size of the stack lowered the frequency at which the loss of vertical control takes place. Figure 4-24A and B shows how double and triple stacks will narrow vertical coverage in the lower frequency range. However, this advantage comes at the expense of a good bit of HF lobing in the vertical plane, due to the multiple sources. Note that the horizontal pattern of these arrays is the same as for a single horn (Figure 4-21A).

If horns are placed side by side, as shown in Figure 4-24C, horizontal coverage will be narrowed, again at the expense of considerable lobing. This solution to narrow horizontal coverage is unsatisfactory, since there are many devices which exhibit consistent narrow coverage

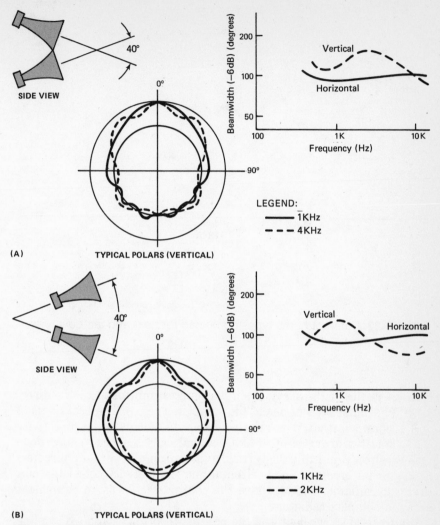

Figure 4-22. Cross-firing of horns. **(A)** Vertical splaying of horns, mouth to mouth. **(B)** Vertical splaying of horns, driver to driver.

angles when used alone. Note here that the vertical pattern of the horizontal array is the same as that of a single horn (Figure 4-21A).

THE PRODUCT THEOREM

Kinsler and Frey[4] present what they call the product theorem, which relates the directivity characteristics of an array of simple (non-directional) sources to a similar array of complex sources. The theorem states that the directional characteristics, as measured in the far field, of the array of complex sources will be the same as the directional

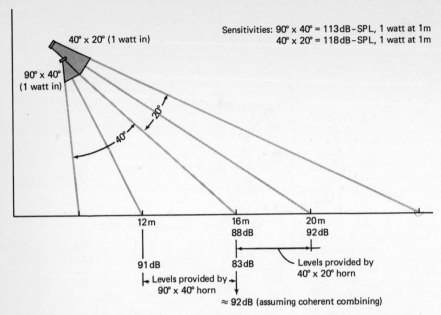

Figure 4-23. Different horns types combined for near and far coverage.

characteristics of the array of simple sources multiplied by the directional characteristics of one of the complex sources. The usefulness of this theorem is that it is relatively easy to calculate the directivity characteristics of arrays of omnidirectional sources. Once we have done this, we simply multiply the vertical and horizontal coverage characteristics of that array by the horizontal and vertical characteristics of a complex radiator, and we have the characteristics of an equivalent array of complex radiators.

Figure 4-25 shows details of the product theorem as applied to a line array of simple sources. The equation in the figure gives the directional characteristics of the line array in the vertical plane as a function of spacing of elements, frequency, and the number of elements. (The array of course will have uniform directional characteristics in the horizontal plane.)

If we substitute directional devices for the omnidirectional ones, then the resulting array will have directional properties in both planes which are the product of the simple array and a single complex device.

The product theorem does not take into account mutual coupling (see Chapter 6) between elements, and this implies that it should be used at wavelengths which are equivalent to the array dimensions, or shorter.

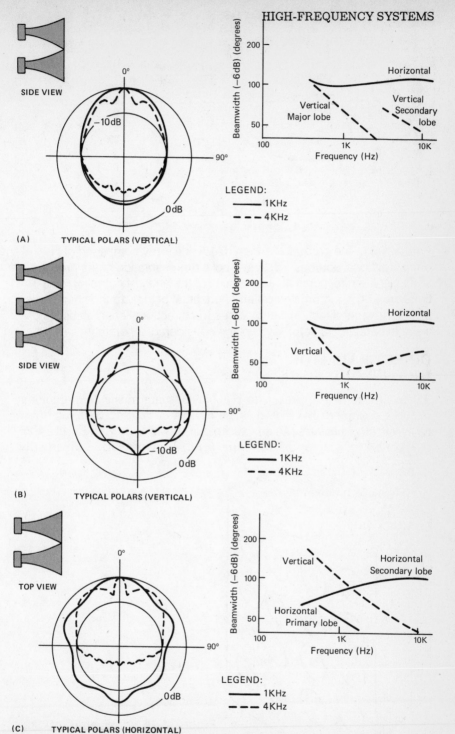

Figure 4-24. Techniques for narrowing coverage. **(A)** A vertical stack of two horns. **(B)** A vertical stack of three horns. **(C)** A horizontal stack of two horns.

93

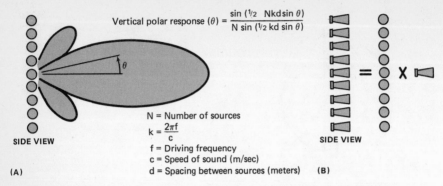

$$\text{Vertical polar response } (\theta) = \frac{\sin (\frac{1}{2} \text{ Nkd} \sin \theta)}{N \sin (\frac{1}{2} \text{ kd} \sin \theta)}$$

N = Number of sources

$k = \dfrac{2\pi f}{c}$

f = Driving frequency

c = Speed of sound (m/sec)

d = Spacing between sources (meters)

SIDE VIEW

SIDE VIEW

(A)

(B)

Figure 4-25. The product theorem. **(A)** A line array composed of N omnidirectional sources. **(B)** A symbolic representation of the product theorem; the directional characteristics of the array of complex radiations are equal, in both vertical and horizontal planes, to the characteristics of the omni-directional array *times* the directional characteristics of a single complex radiator, each plane considered separately.

DIRECT RADIATOR
HIGH-FREQUENCY SYSTEMS

As an alternative approach to HF horns, some music reinforcement systems designers have used large arrays of rugged cone, or direct radiator, loudspeakers. As an example of this, we show, in Figure 4-26, a curved 4-by-4 array of 130 mm (5 in.) loudspeakers. Individually,

Figure 4-26. A HF array of direct radiators.

these loudspeakers can handle 20 watts of program input, and their sensitivity is nominally 94 dB, one watt at one meter. Collectively, through the effects of mutual coupling and increased directivity, one watt feeding the entire array produces 107 dB at a distance of one meter. Since the total power the array can handle is 16 times 20, or 320 watts, this system, fully powered, can produce a level of some 132 dB-SPL at one meter. The nominal directional pattern of the array is 60-by-40 degrees.

By comparison, a typical 60-by-40 constant coverage horn with a 100-watt HF driver will have a sensitivity of 118 dB, one watt at one meter. Fully powered, this combination will produce a level of 138 dB SPL, one watt at one meter.

Thus, the direct radiator array, under many circumstances, may be considered as a reasonable alternative to the conventional horn approach. The large difference, of course, is that the direct radiator array, whatever its maximum output capabilities, will require more power than the horn system. These days, power is relatively inexpensive, and many designers feel that the sonic differences favor the direct radiator array, making the extra expenditure for power well worth it.

VERY-HIGH-FREQUENCY (VHF) COVERAGE: RING RADIATORS

As we have seen, if a compression driver is used on a constant coverage horn, it must be boosted above about 3 kHz in order to provide flat axial and power response. Let us assume that the driver has a power rating of 40 watts. With the boost in the circuit, a 40-watt input signal at 10 kHz will correspond to a maximum input of only 2.5 watts at 1 kHz. This is shown in Figure 4-27A, and it represents a considerable waste of the potential of the 40-watt driver. In essence, we have to apply a mid-band derating to the driver if we wish to feed it full power at 10 kHz.

A far better solution is to operate the 40-watt driver without any HF boost above 3 kHz, allowing it to roll off naturally, and supplementing it with an array of VHF ring radiators.

As shown at Figure 4-27B, the 40-watt driver, operating on a 90-by-40 degree constant coverage horn can produce a maximum level of 129 dB SPL at one meter. On the same graph, we show the maximum level capabilities of one, two, and four ring radiators, each rated at 20 watts input and with a sensitivity of 110 dB, one watt at one meter.

The extra power required to implement an array of ring radiators is minimal, and the approach should be considered whenever a flat program input spectrum is anticipated.

Figure 4-27C shows typical commercial ring radiators.

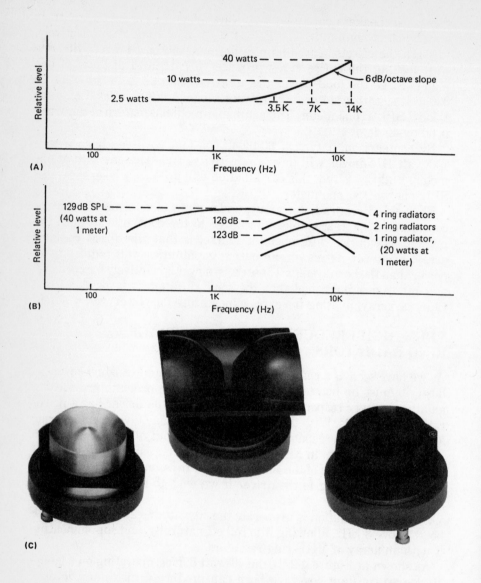

Figure 4-27. Commercial ring radiators. (JBL Photo).

CHAPTER 4:

References:
1. E. Wente and A. Thuras, "Loudspeakers and Microphones for Auditory Perspective," *Electrical Engineering*, pp. 17-24 (Jan 1934).
2. W. Kock and F. Harvey, "Refracting Sound Waves," *J. Acoustical Soc. Am.*, Vol. 21, pp. 471-481 (1949).
3. J. Frane and B. Locanthi, "Theater Loudspeaker System Incorporating an Acoustic Lens Radiator," *J. Soc. Motion Picture and Television Engineers*, Vol. 63, pp. 82-85 (1954).
4. L. Kinsler and A. Frey, *Fundamentals of Acoustics*, John Wiley & Sons, New York (1982).

Additional Reading:
Articles:
1. L. Beranek, "Loudspeakers and Microphones," *J. Acoustical Soc. Am.*, Vol. 26, pp. 618-629 (1954).
2. J. Hilliard, "Historical Review of Horns Used for Audience-type Sound Reproduction," *J. Acoustical Soc. Am.*, Vol. 59, pp. 1-8 (1976).
3. F. Murray and H. Durbin, "Three-dimensional Diaphragm Suspensions for Compression Drivers," *J. Audio Eng. Soc.*, Vol. 28, pp. 720-725 (1980).
4. "Characteristics of High-frequency Compression Drivers," Technical Note Volume 1, Number 8, JBL Incorporated, Northridge, CA (1984).

Low-Frequency Systems

INTRODUCTION

In conventional two-way system design, the LF section extends in frequency from 40-60 Hz at the lower bound up to 500-800 Hz at the upper bound. Early LF systems consisted of folded horns. These devices exhibited fairly high efficiency, which was necessary because of the small amplifiers available at that time. During the forties[1], these horn-loaded systems were modified to become ported horn systems. In these designs, the front of the driver is horn loaded, while the volume behind the driver is vented, providing an extension of LF response down to the 35-40 Hz region. In one form or another, ported horn systems have been a mainstay of the sound reinforcement industry. However, during the mid-seventies, the simple ported enclosure emerged as the preferred LF system for many applications, mainly because of its extended HF power response relative to the older designs, as well as its economy. Sealed LF systems are rarely used in sound reinforcement systems because of their power output limitations at low frequencies.

THE BASIC LF TRANSDUCER

The direct radiator, or cone loudspeaker, is the basis of all LF systems today, and a cutaway diagram of a current model is shown in Figure 5-1A. If a direct radiator loudspeaker is mounted on a large flat baffle, its region of flat power output will be bounded at the low end by the resonance frequency, f_o, of the moving system. The high-frequency bound may be determined by the equation:

$$f_1 = c/2.5d, \tag{5-1}$$

where
f_1 = upper frequency bound,
c = velocity of sound in (m/sec)
d = effective radiating diameter of cone (m).

Figure 5-1B shows curves representing a family of loudspeakers, all of the same diameter, ranging from highly damped (curves 1 and 2) to underdamped (curves 4 and 5). These curves have been normalized; that is, the mid-band, or piston-band portions of their responses, have been made equal.

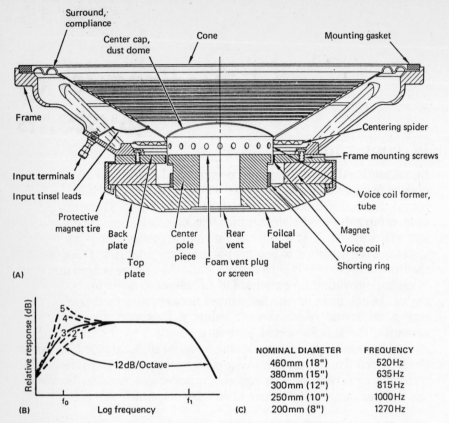

Figure 5-1. Low-frequency drivers.

Magnet size is the dominant factor in providing high loudspeaker damping. The voice coil, moving in the magnetic gap, acts as a potential generator, and that potential is shorted out, or damped, by the low impedance of the amplifier. The greater the magnetic flux density in the gap, the greater the damping. Those loudspeakers with large amounts of electromagnetic damping exhibit rolled-off LF response, but their mid-band sensitivities are quite high.

Depending on its use, the upper frequency bound of a direct radiator may be higher than that indicated by Equation 5-1. This equation gives the frequency above which the power response of the loudspeaker begins to fall off. However, this fall-off is accompanied by an increase in the directivity index of the device, so flat on-axis response can extend beyond f_1. For most applications in sound reinforcement, a useful upper frequency bound may be given by the frequency at which the DI has increased to no more than 6 dB. This value is given by the equation:

$$f = 2c/\pi d, \tag{5-2}$$

where

c = velocity of sound (m/sec), and

d = effective cone diameter (meters)

Figure 5-1C shows the effective upper frequency bounds for typical direct radiator diameters. (In some special designs, such as studio monitor systems, LF transducers may be used up to higher frequencies than those shown here. In these cases, the LF driver is matched to a HF or MF element whose DI may be as high as 10 dB.)

BASIC LF TRANSDUCER CLASSES

In most aspects of sound reinforcement work, we can, within each diameter class, divide LF transducers into three types:
1. Medium sensitivity (Efficiency 0.5 to 3.5)
2. High sensitivity (Efficiency 2 to 6%)
3. Maximum sensitivity (Efficiency 4 to 10%)

These three types are further defined by their applications:
1. *Medium sensitivity:* LF transducers used in high-quality monitor applications in simple ported enclosures. Also used as sub-woofers in special effects applications.
2. *High sensitivity:* LF transducers used in speech and music reinforcement systems. Usually used in ported or in horn loaded ported enclosures.
3. *Maximum sensitivity:* LF transducers primarily used as drivers in LF horn enclosures.

LOUDSPEAKER IMPEDANCES

As shown in Figure 5-2A, the magnitude of impedance of a loudspeaker is a function of frequency. The nominal impedance rating of loudspeakers is usually stated as the minimum value of impedance that occurs above the primary resonance of the moving system. Loudspeakers used in sound reinforcement generally have nominal impedances of 4, 8, and 16 ohms. For a given model of loudspeaker, the nominal impedance can be varied by altering the cross-sectional wire dimensions as shown in Figure 5-2B. Note that in the 16-ohm voice coil, relative to the 8-ohm coil, the edge-wound wire has been flattened to a thickness of 0.7. This allows 1.4 as much wire length to be placed in the gap. However, the cross-sectional area of the wire has decreased 0.7, and the combination of increased wire length and reduced area will double the resistance. Note that the amount of copper in the voice coil has remained constant, as have its dimensions.

When a loudspeaker is placed in a ported enclosure, its impedance magnitude curve shows a characteristic twin peak at low frequencies. An example of this is shown in Figure 5-2C. It occurs as a result of coupling the resonance of the loudspeaker to the resonance of the ported enclosure.

101

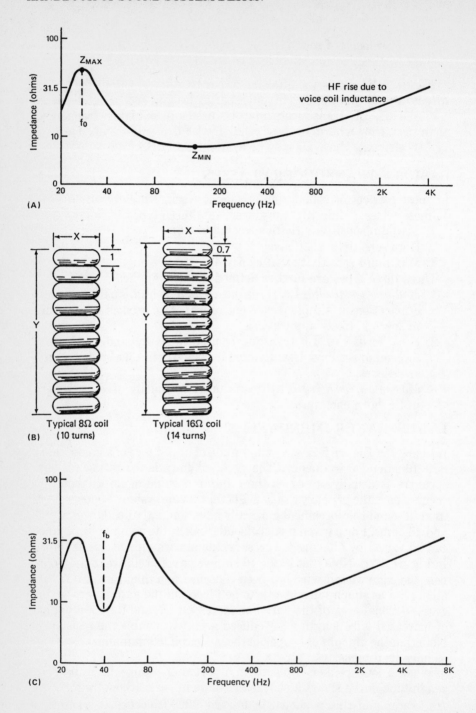

Figure 5-2. Low-frequency transducer impedance.

PORTED LF ENCLOSURES: THIELE-SMALL PARAMETERS

During the fifties, excellent analyses of ported systems were carried out by Beranek[2], Locanthi[3], and Novak[4]. However, it remained for Thiele and Small, elaborating on the work of Novak, to provide a relatively simple means of system syntheses. Their work is richly detailed in the literature[5], and few ported systems are contemplated today which have not first been modelled using these parameters.

There are two coupled resonances in a ported system—the loudspeaker itself and the tuned enclosure; it is the proper matching. or alignment, of the two that will result in a desired system response. In Figure 5-3 note that in the region of enclosure resonance, the actual output from the cone is at a minimum. The enclosure receives power from the loudspeaker in that frequency range, but the cone's motion is slight. Through resonance, the volume-velocity at the port is maximized, and considerable acoustical output results. At resonance, the port output is shifted 90 degrees with respect to the cone. Below resonance, the output of the port approaches 180 degrees relative to the cone, and the response falls off quickly, at 24 dB/octave.

Since the cone's motion is relatively slight in the region of enclosure resonance, the normal displacement non-linearities of the LF transducer are minimized, and the device is usually able to handle its full thermal power rating down to the region of enclosure resonance. Because of system unloading below resonance, it is standard practice in ported systems to high-pass filter the program just below the port resonance frequency.

Thiele and Small identified a set of parameters for LF transducers which enable the designer to calculate the response of a system, with enclosure volume and port tuning as the variables. The parameters useful in this phase of the design are:

Small-signal parameters;

1. f_s, the free air resonance of the transducer, in Hz.
2. Q_{ts}, the total Q of the transducer. The Q referred to here has to do with the sharpness of the displacement resonance curve (See Chapter 1)

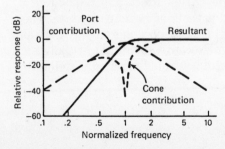

Figure 5-3. Relative output from cone and port in a ported system.

of the transducer when driven by an amplifier with a low source impedance. It is not to be confused with the symbol Q representing the directivity factor of the loudspeaker.

3. V_{AS}, the volume of air providing a restoring force equal to that of the transducer's mechanical compliance, in liters (ft³).

4. η, the half-space reference efficiency of the transducer. This is a measure of the device's conversion efficiency in the midband when it is placed next to one reflective surface, it is given by the following equation:

$$\eta = \frac{4\pi^2 f_S^3 \, V_{AS}}{c \, (Q_{TS})} \tag{5-3}$$

where c is the velocity of sound in air.

Large-signal parameters

1. V_D, the peak displacement volume of the cone (cm³). This is the product of cone area and maximum displacement of the cone from the rest position.

2. x_{max}, peak linear displacement of the cone from rest position (mm).

3. S_d, the effective cone area (m²).

4. $P_{E(max)}$, the thermally limited maximum input power (watts).

5. R_E, the DC resistance of the voice coil.

The small-signal parameters enable the designer to determine the frequency response of a system, while the large-signal parameters enable him to determine its maximum output capabilities.

A Design Example

As an example of how these parameters can be used, let us take a 380 mm (15 in.) LF transducer such as might be used in a studio monitoring system. Its parameters are:

f_S = 20 Hz

Q_{TS} = 0.25

V_{AS} = 460 liters (16.2 ft³)

The alignment we will calculate is a so-called flat alignment; one that will have no ripple or bump in its response as it approaches its lower frequency limits. The approximate design equations, as given by Keele[6], are shortcuts to estimating certain aspects of system response. They have an accuracy of about ± 10% (± 1 dB). A more thorough realization of a vented system program on a microcomputer would actually plot out the relative frequency response of a given simulated system. The equations are:

$$V_b \cong 15 \, (Q_{TS})^{2.87} \times V_{AS} = 129 \text{ liter } (4.5 \text{ ft}^3) \tag{5-4}$$

$$f_3 \cong 0.26 \, (Q_{TS})^{-1.4} \times f_S = 36 \text{ Hz} \tag{5-5}$$

$$f_b \cong 0.42 \, (Q_{TS})^{-0.9} \times f_S = 29 \text{ Hz} \qquad (5\text{-}6)$$

where

V_b = volume of the enclosure

f_3 = the 3-dB down point

f_b = enclosure resonance frequency

The response of this simulated system is shown in Curve A of Figure 5-4.

Let us further calculate the effect of making the box somewhat smaller, say, 85 liters (3 ft³). Again, according to Keele, the 3-dB-down point will be given by:

$$f_3 \cong \sqrt{V_{AS}/V_b} \times f_S = 46 \text{ Hz} \qquad (5\text{-}7)$$

$$H \cong 20 \log [2\text{-}6 \, Q_{TS}(V_{AS}/V_b)^{0.35}] = 1.4 \text{ dB} \qquad (5\text{-}8)$$

$$f_b \cong (V_{AS}/V_b)^{0.32} \times f_S = 34 \text{ Hz} \qquad (5\text{-}9)$$

where

H = the response hump in dB

This humped curve is shown as Curve B in Figure 5-4.

Both kinds of response may be useful. Curve A is obviously the choice for smoothest LF response in a studio monitoring system, while Curve B might be more useful, and convenient because of size, for certain kinds of musical instrument amplification.

The frequency response of the LF system may be determined from the following equation:

$$\text{Response (dB)} = 20 \log \frac{f_n^4}{\sqrt{(f_n^4 - Cf_n^2 + A)^2 + (Bf_n - Df_n^3)^2}} \qquad (5\text{-}10)$$

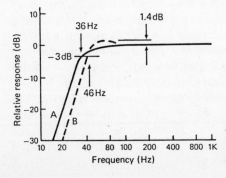

Figure 5-4. Two alignments. Curve A is a natural flat alignment, and Curve B is a "humped" alignment using a smaller enclosure.

105

where:

$$A = \frac{f_b^2}{f_s^2}$$

$$B = \frac{A}{Q_{TS}} + \frac{f_b}{7f_s}$$

$$C = 1 + A + \frac{f_b}{7f_s\,Q_{TS}} + \frac{V_{AS}}{V_b}$$

$$D = \frac{1}{Q_{TS}} + \frac{f_b}{7f_s}$$

At each frequency, f, you wish to calculate, determine $f_n = f/f_s$. Then, substitute each value of f_n into the equation. (Note: This tedious procedure is best carried out with a programmable calculator or a microcomputer.)

Using the large-signal parameters, we will now calculate the displacement limited power rating of the system at the 3-dB-down point:

$$P_{ER} = \frac{(3)\, f_3^4\, V_D^2}{\eta \times 10^{12}} = 221 \text{ watts} \qquad (5\text{-}11)$$

where V_D for the LF transducer we are using is 757 cm^3 and η is .013 (1.3%).

Since the displacement rating is greater than the thermal rating, we know that the system will be limited by the transducer's thermal rating of 150 watts. Had the displacement rating been less than the thermal rating, then *it* would be the limiting factor in power input at f_3.

INTERPRETING MANUFACTURERS' DATA

Today, most manufacturers provide Thiele-Small parameters on all their LF transducers intended for use in ported or horn enclosures. Many manufacturers routinely publish enclosure dimension and port tuning details to enable engineered systems to be built in the field.

Figure 5-5A shows typical values for Thiele-Small parameters for the three major classes of LF transducers in their three most common diameters—460 mm (18 in.), 380 mm (15 in.), and 300 mm (12 in.).

For those who wish to experiment with alternate tunings, the chart shown in Figure 5-5B provides details on port dimensioning. It is always best to use as large a port as is practicable in order to minimize turbulence in the port at the enclosure resonance. Remember, of course, that a port with a fairly large cross-sectional area will generally be quite long, effectively subtracting from the enclosure volume itself.

The data calculated from the Thiele-Small parameters refers only to the LF performance of the system; the HF cutoff will be determined from the information given in Figure 5-1. Further, the analysis assumes that the loudspeakers will be locating adjacent to a single reflective boundary, such as the ground, or a wall.

Figure 5-6 shows examples of typical commercial ported systems. The general range of system sensitivities is from about 93 dB, one watt at one meter (for high linearity monitor systems) to 101 dB, one watt at one meter (for dual LF systems intended for theater use).

CONSTRUCTION OF PORTED SYSTEMS

Most manufacturers of professional quality transducers provide construction plans for those users who wish to build their own enclosures. It is important that these instructions be accurately followed. In general, heavy particle board, generously ribbed, will offer excellent performance. However, the equivalent thickness of plywood will be more rugged for portable applications.

The amount of acoustical damping material inside the enclosure should be minimal—only enough to damp out standing waves at upper frequencies. In fact, the performance of a ported system in the region of enclosure resonance is optimum when there is little or no damping at all, and the design equations presented here make that assumption.

Port diameters for most systems used in professional work should not be less than 100 mm (4 in.) if turbulence in the port at high drive levels near system resonance is to be avoided.

RADIATED POWER VERSUS CONE DISPLACEMENT

The data shown in Figure 5-7 relates acoustical power output to peak cone excursion as a function of frequency and effective cone diameter. Values of x_{max} can be used for given LF transducers in order to determine maximum power output capabilities; however, the chart does not take into account port radiation. As a general approximation, the effective piston diameter for a loudspeaker is its nominal diameter minus 50 mm (2 in.).

Another way to express the relation between cone displacement and radiated power as a function of frequency and cone diameter is given by the following equation:

$$x_{max} = \frac{2\,rp}{\rho_0\,\pi^2 f^2 d^2} \tag{5-12}$$

where

r	= measurement distance (meters)
ρ_0	= density of air (1.21 kg/m³)
d	= piston diameter (meters)
p	= pressure (pascals)
f	= frequency of observation

Figure 5-5. Thiele-Small parameters.

	R_E (Ohms)	f_S (Hz)	Q_{TS}	V_{AS} (liters)	η (percent)	S_D (m²)	X_{MAX} (mm)	V_D (cm³)	$P_{E(MAX)}$ (watts)	L_E (mH)
300 mm (12")										
Medium efficiency	5.7	20	.24	280	0.86	.053	8.0	424	100	0.6
High efficiency	5.5	50	.16	89	6.00	.053	3.5	185	150	1.1
Maximum efficiency	6.3	60	.17	80	8.60	.053	2.5	133	150	0.4
360 mm (15")										
Medium efficiency	6.3	20	.25	460	1.30	.089	8.5	757	150	1.2
High efficiency	6.3	40	.28	175	3.50	.089	5.0	445	200	1.1
Maximum efficiency	5.7	37	.17	300	8.70	.089	3.0	267	100	1.0
480 mm (18")										
Medium efficiency	5.8	20	.27	820	2.10	.130	9.5	1230	300	1.4
High efficiency	6.0	30	.23	480	5.00	.130	5.5	720	300	1.4
Maximum efficiency	6.0	30	.20	425	4.90	.130	5.0	720	300	1.4

(A)

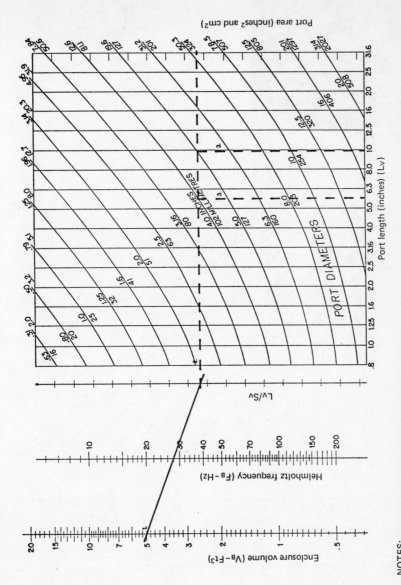

NOTES:
1. This example shows a 5 ft³ enclosure tuned for 28 Hz.
2. From Lv/Sv construction line, draw horizontally for.
3. Port dimension options shown: 4"D x 5½"L and
   5"D x 10"L.

(B)

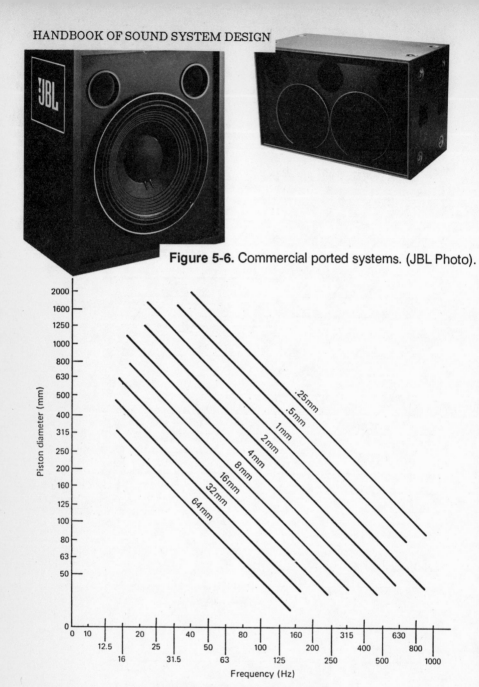

Figure 5-6. Commercial ported systems. (JBL Photo).

For any other power, P, multiply amplitude obtained from chart by $\sqrt{P}$.

EXAMPLE:

   Find peak amplitude in mm for a 400 mm diameter piston radiating 2 watts at 100 Hz.

SOLUTION:

   Find intersection of 100 Hz and 400 mm diameter. Read 0.9.

THEREFORE:

   Peak amplitude = 0.9 $\sqrt{2}$ = 1.26 mm

Figure 5-7. Peak excursion requirements for radiating one acoustic watt on one side of a large flat baffle as a function of piston diameter.

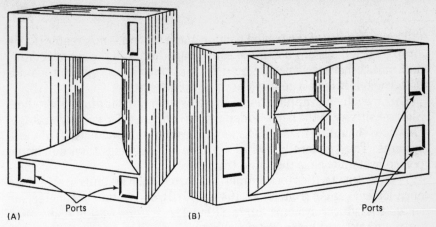

Figure 5-8. Ported horn systems (JBL data).

This expression enables the user to relate cone excursion easily to sound pressure level. The quantity p is in pascals, and it should be remembered that one pascal is equivalent to 94 dB-SPL.

As in the previous case, radiation into half-space or from one side of a large flat baffle is assumed. This equation should be used only when r is large compared to d. The equation further assumes that the radiated wavelength is greater than the circumference of the cone and that radiation is omnidirectional.

PORTED-HORN SYSTEMS

Figure 5-8 shows line diagrams of two typical ported horn enclosures. In one form or another, many manufacturers make systems such as these. The LF transducers are front-loaded with a horn, while behind the transducer there is a volume, tuned by the ports on the front to a frequency in the 40-50 Hz range. The response of these typical systems is shown in Figure 5-9. Note that the horn loading is effective in the double LF system above about 80 Hz, and above about 120 Hz in the single LF system.

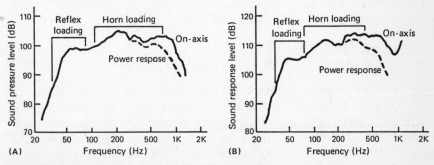

Figure 5-9. Response of ported-horn systems.

The sensitivity of the ported portion of the system is no greater than a simple ported system using the same transducers in the same volume, and unless there is a specific need for the added sensitivity in the mid-bass region, it is best to use a simple ported system.

For speech-only applications, these systems are useful, since their high sensitivity in the horn-loaded region can be used to good advantage. Due note should be made of the power response characteristics of these systems. They generally roll off at lower frequencies than do simple ported systems using the same transducers.

Typical sensitivities for commercial ported-horn systems are, in their horn-loaded range, from about 104 to 106 dB, one watt at one meter, for single transducer systems, up to 110 to 112 dB, one watt at one meter, for the dual transducer system.

LF HORN SYSTEMS

LF horn systems are usually quite sizeable because of the necessity of large mouth dimensions if sufficient LF power is to be radiated. (Recall the discussion of mouth size requirements in Chapter 4.) Klipsch[7] has succeeded in folding horns in an ingenious manner to conserve space normally wasted, and Figure 5-10A shows details of a LF horn designed to work down to 40 Hz. The sensitivity of this system is about 108 dB, one watt at one meter, over its operating frequency range.

Keele[8] has applied the Thiele-Small parameters to LF horn applications, as shown in Figure 5-10B. Here, the various bounds on frequency response are expressed in terms of the relevant parameters. The assumption is made that, at low frequencies, the horn flare rate and mouth size are not the limiting factors. The most significant factor in extending the HF response is to keep the so-called mass breakpoint, f_{HM}, as high as possible. This requires that Q_{ts} be quite low and f_s quite high.

For a typical LF transducer which might be used in a horn system, we would normally expect values of f_{HM} in the 400-600 Hz range. Such drivers, when specifically designed for horn use, are not usually suited to other applications. Their response, when mounted on a flat baffle, would resemble Curves 1 and 2 of Figure 5-1B.

MUTUAL COUPLING;
VERY-LOW-FREQUENCY (VLF) SYSTEMS

In most music reinforcement systems, flat power response down to 40 Hz is felt to be sufficient. However, certain special effects, both in music reinforcement and in the motion picture theater, require extension of the LF bandwidth down to 25 Hz. Because of the ear's relative insensitivity to very low frequencies, considerable amounts of acoustical power must be generated if the effect is to be a convincing one. The

Figure 5-10. Low-frequency horns.

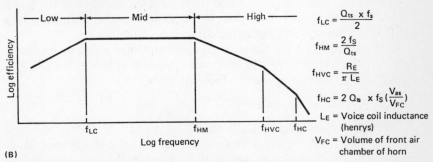

(A)

(B)

$$f_{LC} = \frac{Q_{ts} \times f_s}{2}$$

$$f_{HM} = \frac{2 f_S}{Q_{ts}}$$

$$f_{HVC} = \frac{R_E}{\pi L_E}$$

$$f_{HC} = 2 Q_{ts} \times f_S \left(\frac{V_{as}}{V_{FC}}\right)$$

L_E = Voice coil inductance (henrys)

V_{FC} = Volume of front air chamber of horn

phenomenon of mutual coupling works in our favor by increasing the efficiency of multiple LF radiators over that of a single LF unit.

Figure 5-11A shows the transmission coefficient for a single LF transducer (solid line). When a second transducer of the same type is placed close to it, the two of them, working as a pair, become the equivalent of a new single transducer with 1.4 times the diameter of a single one. The knee in the transmission curve moves down to a frequency 0.7 times the corresponding frequency for the single transducer. When the number of transducers is quadrupled, the knee drops by a factor of 0.5.

113

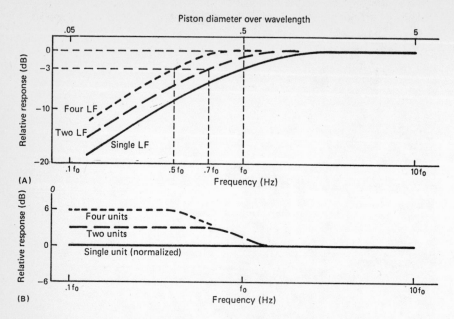

Figure 5-11. Mutual coupling in low-frequency systems.

This set of curves is redrawn normalized at B, and in this form we can see the increase in relative LF output for the same power input.

Remember, however, that when we double the number of LF transducers, we also double the total power input capability, giving us a net 6 dB increase in output capability.

As an example, let us take a single LF transducer rated by its manufacturer at 2% efficiency and 200 watts maximum input. We can now calculate the total acoustical power output:

$$200 \times 0.02 = 4 \text{ acoustic watts}$$

Through mutual coupling, the net efficiency of n number of identical LF drivers will be simply n times the efficiency of a single unit. Thus, for two drivers of the type used in this example, the efficiency will be 4%. Now, let us work our example, assuming that each driver will see 200 watts:

$$400 \times 0.04 = 16 \text{ acoustical watts}$$

In a similar manner, four drivers will result in an efficiency of 8%, and the total power input capability will be 800 watts, yielding a total acoustical power output of 64 watts.

As we increase the number of units in the LF array, the frequency below which we observe mutual coupling continues to move lower, as

114

shown in Figure 5-11. A convenient equation for determining the upper frequency below which the effect is significant is given by:

$$f_{max} = c/d\sqrt{n} \qquad (5\text{-}13)$$

where:

c = velocity of sound (m/sec),
d = nominal spacing between drivers (meters),
n = number of drivers.

Thus, in our examples:

For two drivers:

$$f_{max} = 344/.46\sqrt{2}$$

528 Hz

For four drivers:

$$f_{max} = 344/.46\sqrt{4}$$

374 Hz

The above equation assumes that the LF units are located as closely together as possible, and this is an important requirement in getting the most out of mutual coupling.

Diminishing returns will eventually set in if we try to put too many LF units together. What happens is that f_{max} in the above equation eventually moves down to the region where the LF units are rolling off with successively decreasing frequency, and there is no net gain in efficiency. As a practical matter, it is difficult to get beyond the 15 to 20% range.

Figure 5-12 shows a view of a commercial sub-woofer system.

REVERBERANT FIELD CALCULATIONS

For most indoor calculations, we can use Equation 2-25 to determine actual SPL's in the space. After the actual acoustical power of the combination of loudspeakers has been calculated, it is then entered, along with the room constant, R, for the space under consideration, and the calculation is made. Some manufacturers give published ratings for sub-woofer systems in which the calculation has been made for a reference R. A long-time standard in acoustics has been to refer reverberant levels to a reference room with a room constant of 200 sabins (ft²), which is equivalent to 18.6 SI (metric) sabins (m²).

If such a rating is given, the reverberant level in any new room can be solved by the following equation:

$$\text{New SPL} = \text{Reference SPL} - 10 \log (R/18.6) \qquad (5\text{-}14)$$

where R is the room constant in SI sabins for the new space.

115

Figure 5-12. A commercial sub-woofer system (JBL photo).

DISTORTION IN LF SYSTEMS

If good engineering practice has been applied all the way through the design phase of a sound reinforcement system, then LF distortion should not be a problem. Although all professional grade transducers can stand momentary signal inputs perhaps 6 dB greater than their thermal ratings, care must be taken that displacement limits are not exceeded in the process. A careful analysis of a LF system's performance at its lower frequency bound must be made, and the limits which apply in the frequency range must be applied over the rest of the system's range.

Band-limiting is very important. A horn system must be rolled off at least 12 dB/octave below its effective cutoff frequency. The same applies to ported systems below the enclosure tuning frequency. Ported systems should, if possible, be thermally limited rather than displacement limited in their lower ranges.

The phenomenon known as dynamic compression is shown in Figure 5-13. It is a source of distortion commonly encountered in high-level music reinforcement. The response of a loudspeaker at one-watt input is compared with the response at 100-watts input, and the two curves superimposed. The difference between them is due to physical non-linearities (at low frequencies) and to heating of the voice coil. As the coil heats up, its resistance increases, and the transducer absorbs less power from the amplifier. In a sense, this action represents self-protection of the driver, and as such may not be a bad thing. However,

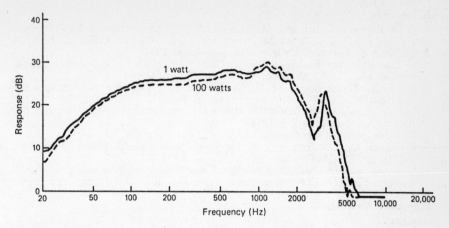

Figure 5-13. An example of dynamic compression in a low-frequency transducer (JBL data).

its effect on music is to rob it of much of its dynamic character. The degree of dynamic compression shown in the figure is fairly slight, and there are many examples which show up to 3 or 4 dB of compression over this 100-to-1 power input range. Copper voice coils exhibit less compression than aluminum does.

USEFUL PARAMETERS FOR LF SYSTEMS

Among the parameters that describe LF systems, the following are felt to be the most useful in laying out systems:

1. Enclosure dimensions, volume, and weight.
2. Transducer complement: model, diameter, and quantity.
3. System impedance: nominal and minimum.
4. Sensitivity: SPL, one watt at one meter.
5. Continuous power rating: watts.
6. Half-space efficiency.
7. Maximum continuous acoustical power output: watts.
8. Maximum continuous SPL: free field at a reference distance and reverberant field level for a reference room constant.
9. Lower frequency limits: –3 dB and –10 dB.
10. Recommended upper crossover frequency.
11. Horizontal and vertical beamwidth (–6 dB); DI, and Q at selected frequencies over the passband of the system.

Data in this form is particularly useful to design engineers in the initial stages of system specification and layout.

CHAPTER 5:

References:

1. J. Lansing and J. Hilliard, "An Improved Loudspeaker System for Theaters," *J. Society of Motion Picture Engineers*, Vol. 45, pp. 339-349 (1945).
2. L. Beranek, *Acoustics*, McGraw-Hill, New York (1954).
3. B. Locanthi, "Application of Electric Circuit Analogues to Loudspeaker Design Problems," *J. Audio Eng. Soc.*, Vol. 19, p. 778 (1971).
4. J. Novak, "Performance of Enclosures for Low-resonance High-compliance Loudspeakers," *IRE Transactions on Audio*, Vol. AU-7, pp. 5-13 (1959).
5. Various, *Loudspeakers* (anthology of articles from the pages of the Journal of the Audio Engineering Society, 1978).
6. D. Keele, personal communication.
7. P. Klipsch, "A Low-frequency Horn of Small Dimensions," *J. Acoustical Soc. Am.*, Vol. 18, pp. 137-144 (1941).
8. D. Keele, "Low-frequency Horn Design Using Thiele-Small Driver Parameters," (presented at AES Convention, Los Angeles, May 1977; preprint number 1250).

Suggested Reading:

Articles:

1. M. Engebretson, "Low-frequency Sound Reproduction," *J. Audio Engineering Soc.*, Vol. 32, pp. 340-346 (1984).
2. M. Gander, "Moving Coil Loudspeaker Topology as an Indicator of Linear Excursion Capability," *J. Audio Eng. Soc.*, Vol. 29, pp. 10-26 (1981).
3. M. Gander, "Dynamic Linearity and Power Compression in Moving-coil Loudspeakers." (presented at AES Convention, New York, 8-11 October 1984; preprint number 2128).
4. H. Harwood, "Loudspeaker Distortion Associated with Low-frequency Signals," *J. Audio Eng. Soc.*, Vol. 20, pp. 718-728 (1972).
5. C. Henricksen, "Heat Transfer Mechanisms in Moving Coil Loudspeakers," (presented at AES Convention, Los Angeles, 10-13 May 1977; preprint number 1247).
6. "Distortion and Power Compression in Low-frequency Transducers," Technical Note Volume 1, Number 9, JBL Incorporated, Northridge, CA (1985).
7. G. Margolis and R. Small, "Personal Calculator Programs for Approximate Vented-box and Closed-box Loudspeaker System Design," *J. Audio Eng. Soc.*, Vol. 29, pp. 421-441 (1981).

Special Mid-Frequency Systems

INTRODUCTION

As we have stated before, sound reinforcement systems have traditionally been of two-way design, with the transition between LF and HF taking place in the 0.5 to 1 kHz range. In the last twelve or fifteen years, the requirements for high-level music reinforcement have pointed up certain weaknesses in the two-way approach. Specifically, HF systems have clear power handling limitations in the 0.5 to 1.2 kHz range, and LF systems are apt to exhibit rolled-off power response in the 0.5 to 1 kHz range.

Originally, theater systems covered the range from about 50 Hz to about 5 kHz. It was natural to divide the spectrum at 500 Hz, which is the *geometric mean* between the frequency range extremes. This frequency is given by the equation:

$$f_o = \sqrt{(f_l) \times (f_h)} \tag{6-1}$$

As is shown in Figure 6-1A, the transition frequency of 500 Hz results in equal distribution of power on an octave basis, assuming a flat input spectrum.

If we extend the overall system bandwidth from 40 Hz to 20 kHz, which is typical of many reinforcement and monitor systems, we can conceivably divide the spectrum into three bands, as shown in Figure 6-1B. Note that each of these three bands covers an equal portion of the spectrum on an octave basis. The range between 315 Hz and 2.5 kHz is now a likely candidate for separate coverage, filling in as it does for the power response and distortion problems inherent in that range in conventional two-way systems.

SPECIFIC HARDWARE

Compression Driver Systems

Those manufacturers who have addressed this problem have generally chosen the decade between 200 Hz and 2 kHz as their target. Figure 6-2 shows an example of a mid-frequency (MF) compression driver system designed to handle this decade. This design represents a scaling down of the traditional HF compression driver, as discussed in Chapter 4. The throat exit area and the phasing plug slit areas have been significantly increased so that distortion due to high

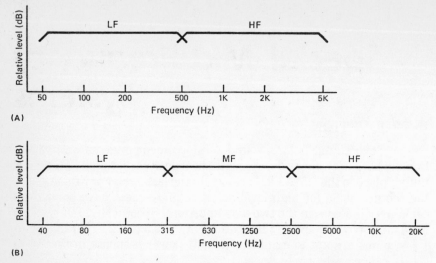

Figure 6-1. Frequency division in two- and three-way systems. **(A)** An early two-way theater system. **(B)** A modern three-way system.

acoustic intensities in the throat area is minimized. With their more robust voice coil structures, these devices can handle more input power than their HF counterparts, and efficiencies in the 25-30% range are typical. These units can deliver acoustical power outputs approaching 100 watts. New families of horns are of course required to complement these devices.

Direct Radiator Systems

As was the case with HF systems, direct radiators can be used in multiples to cover the 200 Hz to 2 kHz range with reasonable sensitivity and power handling. Figure 3 shows an array of nine 300 mm (12 in) loudspeakers used for this purpose. The drivers used here are ordinarily of the sort used for musical instrument amplification, and over the bandwidth employed here, these drivers would have input power ratings of about 150 watts. The sensitivity of a single driver would be about 103 dB, one watt at one meter, while the total array, due to increased directivity, would have a sensitivity of about 108 dB, one watt at one meter. Since the entire array of nine drivers can handle 1350 watts input, the maximum acoustical output pressure level will be:

108 + 10 log (1350) = 139 dB, referred to one meter

(Note that we would not actually measure a level of 139 dB at a distance of one meter, since that is well in the near field of the array. Using this sensitivity as a reference, we would typically measure a level of

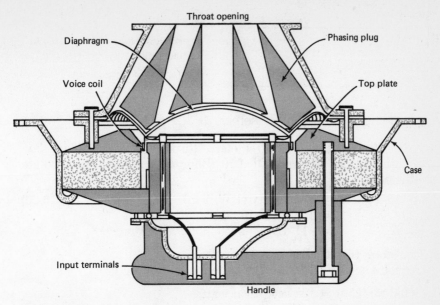

Figure 6-2. An MF compression driver system. A drawing of the Community Light&Sound M4 system.

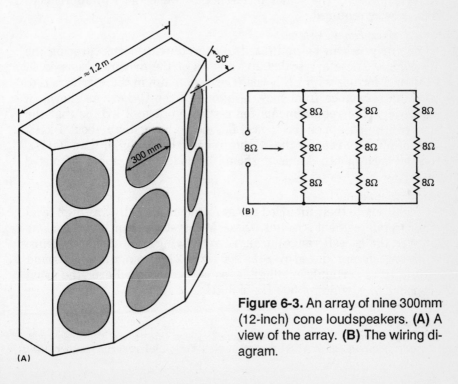

Figure 6-3. An array of nine 300mm (12-inch) cone loudspeakers. **(A)** A view of the array. **(B)** The wiring diagram.

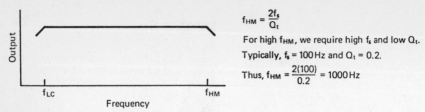

Figure 6-4. Cone drivers of MF horns.

119 dB at a distance of 10 meters, through inverse square law attenuation.) If we assume a DI of 10 dB for the array, we can calculate its efficiency:

$$10 \log (\text{eff}) = \text{Sensitivity (1 W, 1 m)} - \text{DI} - 109 \qquad (6\text{-}20)$$
$$= 108 - 10 - 109$$
$$= -11 \text{ dB}$$

Therefore, the efficiency is .08, or 8%.
The total acoustical power the system can produce is:

$$(.08) \times (1350) = 108 \text{ acoustical watts.}$$

This is a rather substantial amount of acoustical power, and it is essentially equal to the output capability afforded by the compression driver approach. The major difference between the two approaches is the power required.

Cone Driver Horn Systems

Cone drivers can be optimized to operate with horns covering the MF range. These are scaled-up versions of the drivers discussed in Chapter 5 for use with LF horns. The design aim in these drivers is to keep the resonance frequency, f_s, high and the Q_t low. As shown in Figure 6-4, the requirements for response out to 2 kHz is that the mass-controlled upper frequency, f_{HM}, be no lower than about 1 kHz. The cut-off flare rate for the MF horn should be no higher than 100 Hz, and the horn mouth diameter should be no less than about 0.5 meter (20 in).

Final Comments

In addition to their intended use as elements in multi-way high-level music reinforcement systems, these MF systems can be used alone to cover the speech range in high-level paging or warning systems. While speech reproduced over the 200 Hz or 2 kHz range may not sound natural, it is certainly intelligible, as is speech heard over the usual telephone bandwidth of 300 Hz to 3 kHz.

CHAPTER 6:
Suggested Reading:

1. B. Howze and C. Henricksen, "A High-efficiency, One-decade Midrange Loudspeaker," (presented at AES Convention, New York, 1981; preprint number 1848).

Dividing Networks and Matching of Components

INTRODUCTION

Thus far, we have discussed the basic components of sound reinforcement systems. We are now ready to combine them into systems tailored for specific jobs. Traditionally, LF and HF elements are coupled via a passive dividing network to form a complete system. For many applications, especially those requiring moderate output levels, this approach is reasonable. In recent years, however, there has been a major trend toward *biamplification* and *triamplification*. In these systems, separate amplifiers are used to drive either two or three elements, with frequency division taking place before the amplifiers. The basic approaches are shown in Figure 7-1.

In this chapter, we will discuss how HF and LF systems are combined by the user or sound contractor in typical field applications, and the advantages of bi- and triamplification will be discussed in detail.

PROCEDURE FOR COMBINING HF AND LF ELEMENTS

When HF and LF elements are to be combined by means of a high-level, or passive, network, as shown in Figure 7-1A, the following rules should be observed:

1. Establish the LF coverage requirements, noting carefully the number of elements required to do this.

2. Establish the HF coverage requirements for the system.

3. Note the sensitivity difference between the LF and HF elements and ensure that the network can accommodate this difference. Ensure further that the HF portion of the system has been chosen to have sufficient power handling capability so that full power can be passed through the network to the LF part of the system. In other words, the HF portion of the system should not be the limiting factor in powering the entire system.

4. Finally, ensure that the two elements are placed together in such a way that they combine properly in the region of crossover.

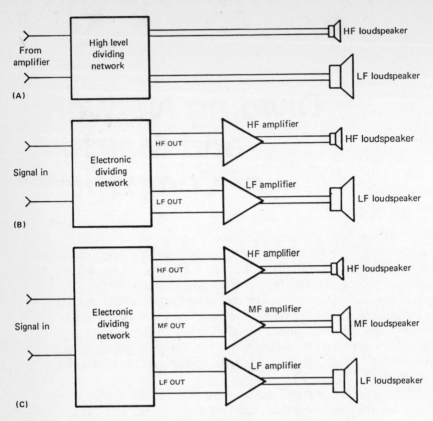

Figure 7-1. Single and multiple amplification of loudspeakers.
(A) A high-level, passive dividing network. **(B)** An illustration of bi-amplification. **(C)** An illustration of tri-amplification.

Figure 7-2A shows details of a typical 12 dB/octave network used in sound reinforcement. At B we show a photo of a typical network, and at C we show response curves for various HF settings of the network.

The purpose of the network is to channel portions of the input signal to those elements that can reproduce them. Additionally, the network protects the HF components by keeping large LF signals out of them. 12 dB/octave networks are common in sound reinforcement because the network slopes are sharp enough to provide reasonable HF component protection. In Figure 7-3, we illustrate the phase relationships between HF and LF outputs of the network. Note that either the HF or LF transducer must be connected in reverse polarity if the signals are to add properly at crossover.

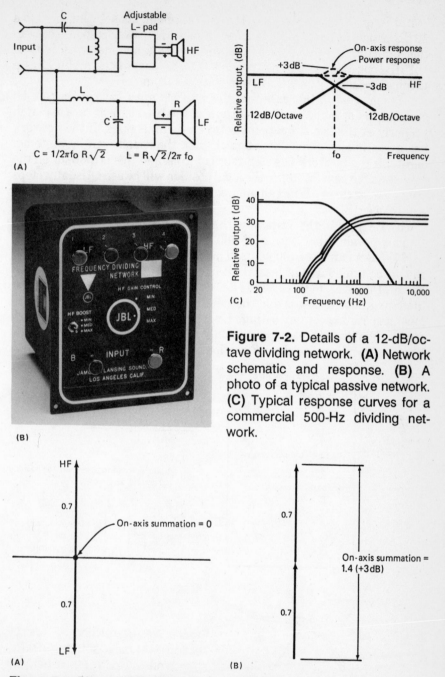

(A)

$C = 1/2\pi f_o R \sqrt{2}$ $L = R\sqrt{2}/2\pi f_o$

(B)

(C)

Figure 7-2. Details of a 12-dB/octave dividing network. **(A)** Network schematic and response. **(B)** A photo of a typical passive network. **(C)** Typical response curves for a commercial 500-Hz dividing network.

(A)

(B)

Figure 7-3. Phase relationships in a 12-dB/octave dividing network. **(A)** Normal connection. **(B)** Reversed connection.

In Figure 7-4, we show how HF and LF elements are connected to a 12 dB/octave network which has additional HF power response correction built in. The LF part of the system has a nominal sensitivity of 101 dB, one watt at one meter, and the HF part of the system has a nominal sensitivity of 112 dB, one watt at one meter.

It is apparent in this case that the network must have at least 11 dB of mid-band attenuation if the outputs of the two elements are to match properly at the transition point between them, 500 Hz. If the network has a continuously variable control, then the match can be quite accurate. If the network has step-type controls, and this is more often the case, then the nearest step to the ideal position will be used. Usually, steps are in two or three dB increments, so it should be possible to get a match within one dB.

The LF part of the system, with its two LF transducers, has a continuous power input rating of 400 watts, while the HF compression driver has a rating of 100 watts. This tells us that a 400-watt amplifier should be attenuated at least 6 dB in order to avoid damage to the HF driver. However, in this case, 11 dB of attenuation is required for proper matching, which means that the HF driver will always receive a signal less than its maximum rating. Under this condition of HF network attenuation, the HF element will only receive about 30 watts when 400 watts are available to the LF section.

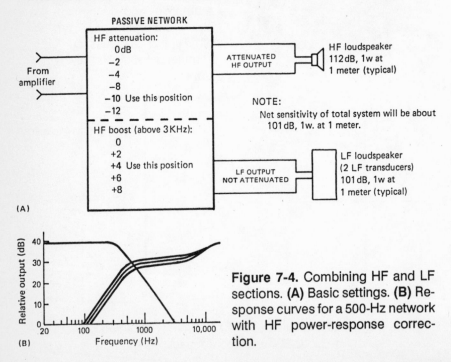

Figure 7-4. Combining HF and LF sections. **(A)** Basic settings. **(B)** Response curves for a 500-Hz network with HF power-response correction.

126

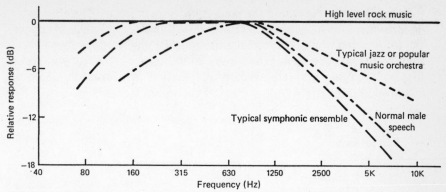

Figure 7-5. Typical spectra (long-term average) for various types of program material.

If we are using a constant coverage HF horn, then some degree of HF boost will be desirable in order to correct for the natural power response roll-off of the HF driver. Obviously, this boosting will result in the feeding of more power to the HF driver, but in this system there is an adequate margin. In fact, the margin is the difference between 30 watts and 100 watts, some 5 dB. Thus, we should be able to insert up to 5 dB of power response correction through the network.

Often, there will not be such a comfortable margin, and the user must run some risk in adding power response correction. Figure 7-5 shows the normal spectra for various kinds of program material. Note that the spectra for normal male speech, classical music, and most kinds of popular music tend to be rolled off. Thus, with only a little care, HF elements can be provided with excess power input capability beyond their maximum rating if they are used in reinforcing rolled-off spectra. Of course, for rock reinforcement applications, care must be taken that there are sufficient HF drivers to allow for a continuous flat spectrum at the system's output.

PHYSICAL PLACEMENT OF HF AND LF ELEMENTS

In general, manufacturers' recommendations for physical placement of HF and LF sections of a system should be followed. At all costs, a cancellation in output in the crossover region must be avoided. Such problems can be detected by feeding a sine wave signal to the system at the crossover frequency and observing the output on axis with a sound level meter. If a reversal of the leads to the HF transducer results in increased output, then the system should be operated that way. However, care must be exercised to ensure that all other elements in the system are compatibly poled.

In many cases, reversing the polarity will show little difference in the on-axis output. The reason for this is shown in Figure 7-6A. Because of the orientation of the HF section, the relative phase angle at crossover may be in, say, the 90-degree range. Switching the polarity will result in –90 degrees, and the vector summation will remain of equal magnitude.

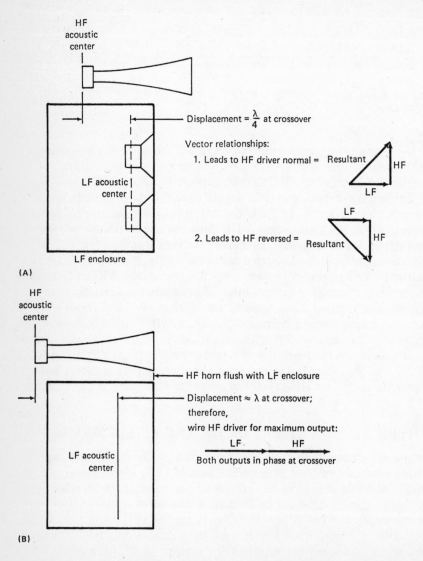

Figure 7-6. The HF and LF element physical relationships. **(A)** Poling of HF unit. **(B)** Mounting the HF horn flush with the LF enclosure.

128

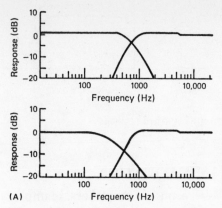

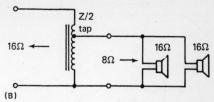

Figure 7-7. The loading of passive dividing networks. **(A)** Properly and improperly matched driver/network combinations. **(B)** Load impedance correction with an autotransformer.

A good rule to follow is this: mount the HF horn so that its mouth is flush with the upper edge of the LF enclosure. Then, make the polarity test and determine which poling, if either, results in the greatest output. If both positions produce about the same output, then use the manufacturer's recommended wiring convention.

Details of this are shown in Figure 7-6B. The advantages of keeping the HF horn mouth flush with the front of the LF enclosure are minimized interferences in the crossover region and smoother vertical polar response in the crossover range.

EFFECT OF DRIVER IMPEDANCES AND NATURAL ROLL-OFFS ON CROSSOVER FREQUENCY

High-level, passive dividing networks must be properly loaded if they are to work properly. At Figure 7-7A, we illustrate the results of proper and improper loading on a passive network. If need be, autotransformers should be used to correct load impedances; an example of this is shown at B.

LF transducers exhibit a natural roll-off at their upper limit, and the slope of this roll-off is about 12 dB/octave. If the LF elements are crossed over in this range, then the actual crossover slope will be the *sum* of that of the driver and that produced by the network. An example of this is shown in Figure 7-8. This is an additional factor in the variability of HF and LF summation in the crossover region, and it is one more reason why measurements of system response should be made before final poling is settled. It is also one more reason why manufacturers' recommendations should be followed.

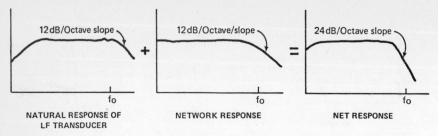

Figure 7-8. Combining network slopes and natural roll-offs.

BIAMPLIFICATION AND TRIAMPLIFICATION

Due to the general availability of stereophonic amplifiers, biampli-fication, or "biamping," has become quite common. Its advantages over a single amplifier are:

1. *Reduced Distortion at High Drive Levels.* In a conventional system, LF overload results in the production of higher harmonics. These distortion components are passed through the dividing network into the HF part of the system. In a biamped system, this cannot happen.

2. *Improved Loudspeaker-Amplifier Damping Factor.* Damping factor is the ratio of the loudspeaker load to the total impedance seen by the loudspeaker looking back into the amplifier. It is given by the following equation:

$$\text{Damping Factor} = Z_L/(Z_O + R_1), \dots\dots\dots\dots\dots\dots\dots(7\text{-}1)$$

where Z_L is the loudspeaker impedance, Z_O the nominal source imped-ance of the amplifier, and R_1 the line resistance between amplifier and loudspeaker. The removal of inductances from the LF part of the passive network improves damping factor and results in smoother response.

3. *More Effective Allocation of Available Power.* In a biamped system, both HF and LF sections may be accurately matched with their rated drive power. In the example we have just worked, a combination of 400 watts for the LF section and 100 watts for the HF section will result in about 3 dB more output capability—with the same safety margin—than if a single 400-watt amplifier with a passive network were used.

Figure 7-9A shows typical schematics for biamping. B shows photos of standard low-level, or electronic, dividing networks. Some of these networks have variable crossover points and variable slopes, while others, using plug-in cards, have fixed values. Note that DC protection capacitors should always be used in series with the HF driver in biamped systems in order to protect the HF driver from turn-on or turn-

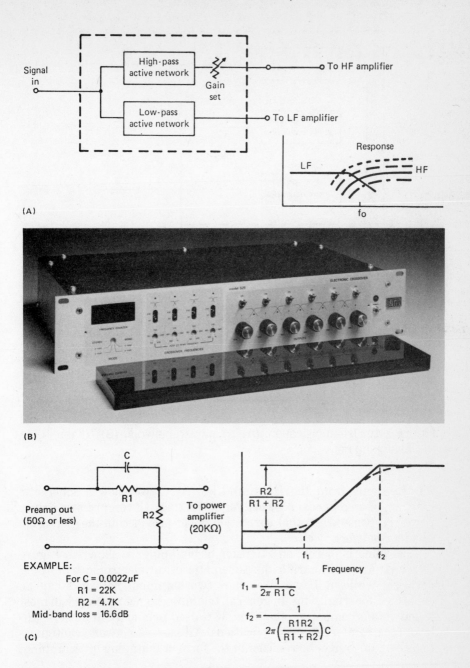

(A)

(B)

EXAMPLE:
For C = 0.0022 μF
R1 = 22K
R2 = 4.7K
Mid-band loss = 16.6 dB

(C)

$$f_1 = \frac{1}{2\pi \; R1 \; C}$$

$$f_2 = \frac{1}{2\pi \left(\dfrac{R1R2}{R1 + R2}\right) C}$$

Figure 7-9. Bi-amplification. **(A)** Bi-amping network. **(B)** Photos of commercial networks for bi-amping. **(C)** A circuit for HF compression driver power response correction.

131

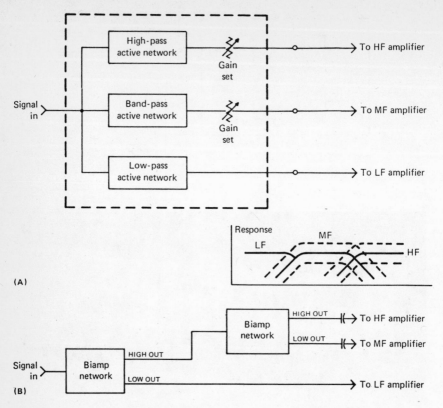

Figure 7-10. Tri-amplification. **(A)** Tri-amping network. **(B)** Tri-amping via two bi-amping networks.

off transients from the HF amplifier. At C, we show a schematic diagram of an external network for correcting HF compression driver power response when that correction is not provided in the electronic dividing network itself.

Triamping is merely an extension of biamping, as shown in Figure 7-10. At A, we illustrate the schematic of an electronic triamping network, while at B we show how two biamping networks can be connected for triamping. In general, triamping is not necessary in most sound reinforcement applications. It is used primarily in large, high-level rock reinforcement applications. All passive network configurations can, of course, be realized in the form of biamping or triamping.

HIGHER-ORDER NETWORKS

While 12 dB/octave networks are adequate for most sound reinforcement applications, there are times when *higher order* networks are desirable. In this sense, the term *order* refers to the number of elements in the network that determine the slope:

Order	Slope	Number of L or C Elements that determine roll-off slope
1st	6 dB/octave	1
2nd	12 dB/octave	2
3rd	18 dB/octave	3
4th	24 dB/octave	4

First order networks are not used in sound reinforcement because they do not afford adequate out-of-band protection for HF drivers. Third order networks may be used in high-level applications where their rapid slopes provide greater protection for HF drivers. Figure 7-11 shows details of a third-order network. Note that the total phase shift through the system (but not its amplitude response) depends on the polarity of one of the elements.

FOURTH-ORDER NETWORKS

A special form of this network, described by Linkwitz[1], has excellent overall characteristics. While most network designs result in HF and LF attenuation of 3 dB at crossover, this network produces HF and LF attenuation of 6 dB at crossover. This results in a flat on-axis summation at crossover, since both HF and LF outputs are in phase at

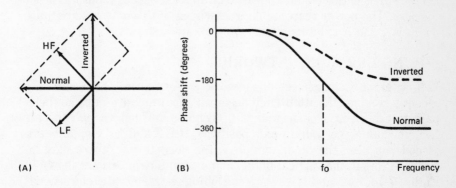

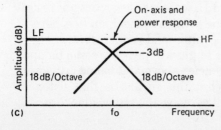

Figure 7-11. Details of a third-order (18-dB/octave network. **(A)** Phase relationships at crossover. **(B)** Phase shift, normal and inverted connection. **(C)** Frequency response.

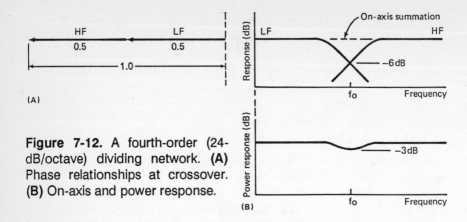

Figure 7-12. A fourth-order (24-dB/octave) dividing network. **(A)** Phase relationships at crossover. **(B)** On-axis and power response.

that frequency. There is a negligible 3-dB drop in power response at crossover, but its bandwidth is extremely narrow, due to the quite rapid slopes. Details of this network are shown in Figure 7-12.

CONSTANT VOLTAGE NETWORKS

Figure 7-13 shows the schematic of what may be called a constant voltage network. Regardless of the high-pass slope, the summation of HF and LF on-axis output will always be the same as the input signal. On the surface, this would appear to be an ideal network, but this is not the case. The power response characteristics of this network are quite irregular, and it exhibits gross *lobing* error.

LOBING ERROR IN NETWORKS

Lobing error is an aspect of network performance which has only recently received the attention it deserves[2]. Too much attention has been paid to the on-axis performance of systems and not enough to performance along certain off-axis positions. If HF and LF elements are placed one over the other, which is the normal case, there may be lobing errors in the vertical plane. Some examples of this are shown in Figure 7-14. At A and B we show the lobing characteristics of a second-order network at crossover. At C and D, we show the errors above and below crossover. When the polar lobe is symmetrical, as it is here at crossover, the lobing error is said to be minimal. Listeners located above or below the major axis by the same angle will hear the same balance from the two components. At frequencies above and below crossover, the error remains symmetrical. What this means is that listeners above the normal on-axis position will perceive the same balance as will listeners below the axis.

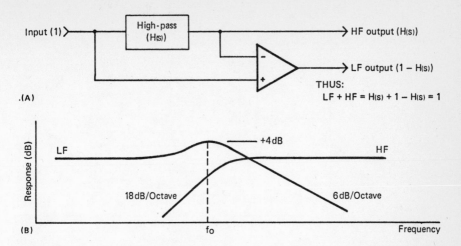

Figure 7-13. Constant-voltage dividing networks. **(A)** The principle of the constant potential network. **(B)** HF and LF frequency response.

At E, F, and G, we show the lobing errors in a third-order system. At crossover there is considerable lobing error, while at frequencies above and below the error is small, due to the rapid slopes.

At H, I, and J, we show the lobing errors for a fourth-order network. Note that lobing error is minimal in all cases.

For most applications in sound reinforcement, lobing errors may not be of major importance. However, in monitoring systems and other high-quality playback applications, the consideration is an important one.

Figure 7-14. Lobing errors in vertical polar response; HF and LF elements are separated by one wavelength at 1 kHz. **(A)** Second-order, normal connection; note cancellation on axis. **(B)** Second-order, reversed connection. **(C)** Second-order, one-half octave above crossover. **(D)** Second-order, one-half octave below crossover. **(E)** Third-order, reversed connection; note that major lobe points downward. **(F)** Third-order, one-half octave above crossover. **(G)** Third-order, one-half octave below resonance. **(H)** Fourth-order. **(I)** Fourth-order, one-half octave above crossover. **(J)** Fourth-order, one-half octave below crossover.

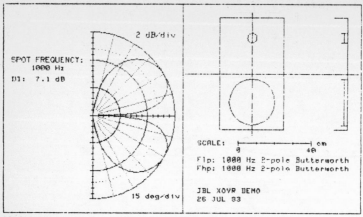

(A) SECOND-ORDER, NORMAL CONNECTION. NOTE CANCELLATION ON AXIS.

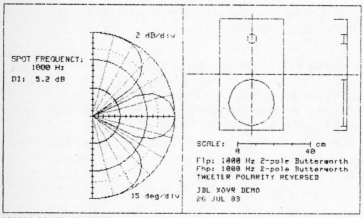

(B) SECOND-ORDER, REVERSED CONNECTION.

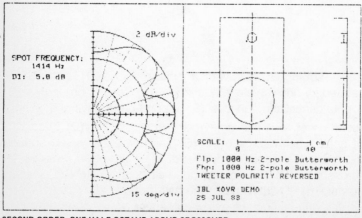

(C) SECOND-ORDER, ONE-HALF OCTAVE ABOVE CROSSOVER.

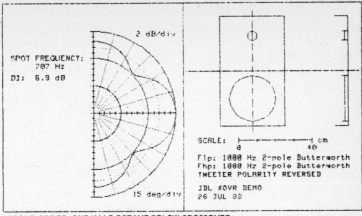

(D) SECOND-ORDER, ONE-HALF OCTAVE BELOW CROSSOVER.

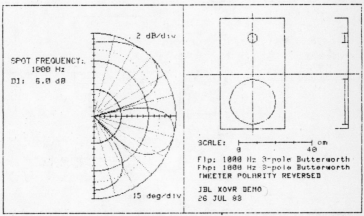

(E) THIRD-ORDER (REVERSED CONNECTION). NOTE THAT MAJOR LOBE POINTS DOWNWARD.

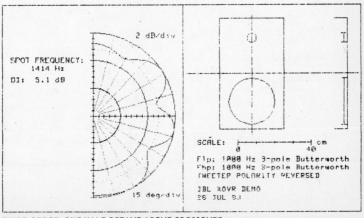

(F) THIRD-ORDER, ONE-HALF OCTAVE ABOVE CROSSOVER.

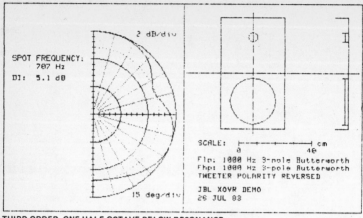

(G) THIRD-ORDER, ONE-HALF OCTAVE BELOW RESONANCE.

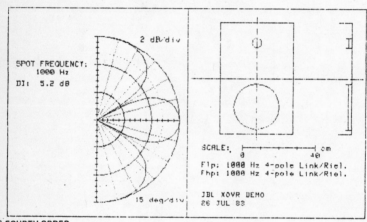

(H) FOURTH-ORDER.

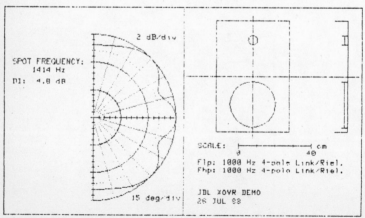

(I) FOURTH-ORDER, ONE-HALF OCTAVE ABOVE CROSSOVER.

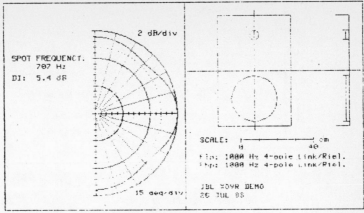

SPOT FREQUENCY.
707 Hz

DI: 5.4 dB

2 dB/div

15 deg/div

SCALE:
0   40 cm

+ln: 1000 Hz 4-pole Link/Riel.
+hp: 1000 Hz 4-pole Link/Riel.

JBL MOVR DEMO
26 JUL 85

(J) FOURTH-ORDER, ONE-HALF OCTAVE BELOW CROSSOVER.

TIME DOMAIN PERFORMANCE OF SYSTEMS

Time domain considerations have to do with the relative time delays between sections of a multi-way system. Ideally, all sections should be "in step," but to accomplish this entails considerable network componentry.

In addition to the delay characteristics inherent in a given driver (because of its band-pass characteristics), networks add their own delay. Suitable stepping of the baffle may correct time offset problems between components, but at the same time there may be adverse reflections from the stepped surfaces. Electronic or passive delay lines can be used to correct problems, but these approaches are quite expensive.

Figure 7-15A shows the basic problem as it exists in a small three-way system composed of cone-type drivers. At B, the stepped baffle has corrected the time offset problem, but there is now the possibility of adverse reflection effects off the steps in the baffle. A gradual slope in the baffle, as shown at C, is a better solution.

Corrections of the sort we have shown here may not be worth the trouble, when one considers that sound reinforcement systems are usually listened to over a very wide vertical angle. A correction made along one vertical angle may very well result in an error along another direction.

More fundamentally, we must ask the question of just how audible time offsets really are. Most of the immediate audibility of variable time delay in portions of a multi-way system *does not* result from the effects of relative time delay itself, but rather from amplitude variations in the crossover region due to delay-dependent reinforcements and cancellations between drivers. When these effects are held constant,

139

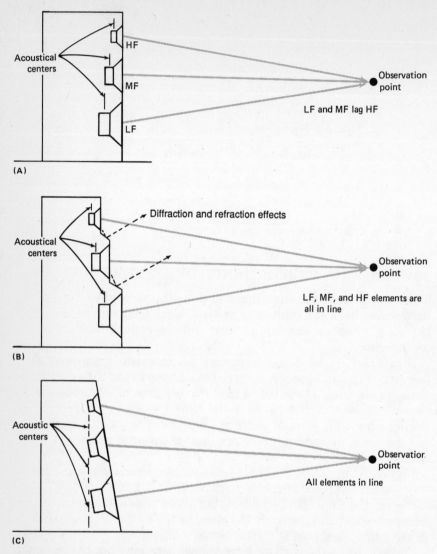

Figure 7-15. Time domain relationships in loudspeaker systems. **(A)** A conventional vertical flat baffle. **(B)** A stepped baffle. **(C)** A sloped baffle.

and delay is introduced *externally* to the loudspeaker system, then the delay effects are much less noticeable.

Blauert and Laws[3] have studied the phenomenon in detail, and the data of Figure 7-16 summarizes their findings. Given very critical listening conditions, the threshold of audibility of delay effects is as shown in the figure. Listeners are most sensitive in the 2 kHz region,

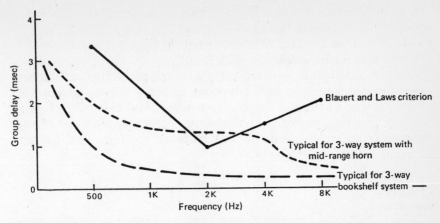

Figure 7-16. Blauert & Laws data on the audibility of time offset.

and this indicates that component placement should be carefully considered if crossovers are to be made in this region. The relatively long HF horns provide most of the delay offset in sound reinforcement systems; this indicates that the LF sections should be delayed with respect to them for proper alignment. These effects are generally inaudible, and their correction in routine sound reinforcement work is not worth the trouble. A far more important consideration is that of maintaining smooth amplitude response through the transition between adjacent elements.

WHICH NETWORK TYPES TO USE

The choice of network types depends on a number of considerations; therefore, we will present a summary of the various orders. In this discussion, on-axis response refers to the acoustical vector summation of the signals for adjacent drivers as measured in the direct field. The power response of the system is proportional to the squares of the acoustical pressures produced by adjacent drivers, and it is a measure of the power output of the system in the reverberant field.

1. *First-order networks.* Not used in sound reinforcement because the gentle slopes do not provide sufficient protection for HF components. Both on-axis and power response are flat through the transition region. Lobing error is moderate.

2. *Second-order networks.* For flat power response through the crossover region, there is a 3-dB increase in on-axis response. Lobing error is minimal, and HF component protection is usually adequate.

3. *Third-order networks.* Provides added HF protection due to rapid slopes. Both on-axis and power response are flat through the crossover region, but lobing error is moderately severe.

141

4. *Fourth-order networks.* For flat on-axis response, the power response exhibits a 3-dB dip. However, due to the quite rapid slopes, the bandwidth of the dip is negligible. Sharp slopes provide excellent HF protection, and lobing error is quite small.

5. *Constant-voltage networks.* On-axis response for a listener precisely positioned is smooth, but power response is quite erratic over a large range. Further, a listener even slightly off-axis will perceive erratic direct field response. Lobing error is probably the worst of any network type, and this kind of network is not recommended for any aspect of sound reinforcement.

CHAPTER 7:

References:

1. S. Linkwitz, "Active Crossover Networks for Noncoincident Drivers," *J. Audio Eng. Soc.*, Vol. 24, pp. 2-8 (1976).
2. R. Bullock, "Satisfying Loudspeaker Crossover Constraints with Conventional Networks," *J. Audio Eng. Soc.*, Vol. 31, pp. 489-499 (1983).
3. J. Blauert and P. Laws, "Group Delay Distortion in Electroacoustical Systems," *J. Acoustical Soc. Am.*, Vol. 63, No. 5 (1978).

Additional Reading:

Articles:

1. G. Augspurger, "Versatile Low-level Crossover Networks," *db Magazine*, (March 1975).
2. G. Augspurger, "The Importance of Speaker Efficiency," *Electronics World* (January 1962).

Books:

1. J. Eargle and G. Augspurger, *Sound System Design Reference Manual*, JBL Incorporated, Northridge, CA (1986).
2. Various, *Loudspeakrs*, (Volumes 1 and 2; compiled from the pages of the Journal of the Audio Engineering Society, New York, 1978 and 1984).

Microphones in Sound Reinforcement

INTRODUCTION

Since there is such a wealth of information on microphone design and application, our coverage in this chapter will deal only with those aspects important in sound reinforcement. After a brief coverage of the basic principles of transduction, we will cover the basic pickup patterns and how those patterns are used in sound reinforcement. We will then discuss proximity effect at low frequencies and beaming at high frequencies. Special microphone types will be discussed, as well as interference effects between microphones. Finally, we will cover microphone specifications and various electrical considerations, such as remote powering of capacitor types, line losses, and polarity conventions.

BASIC PRINCIPLES OF TRANSDUCTION

Essentially, there are two types of transducers used in microphones today: the *capacitor*, or condenser, type and the *dynamic*, or moving coil type.

Capacitor microphones:

The capacitor principle is shown in Figure 8-1A. A capacitor consists of two closely spaced metal plates, and for a given applied potential, E, there will be a charge, Q, stored in the capacitor such that:

$$E = Q/C, \tag{8-1}$$

where Q is the charge in coulombs and C the capacitance in farads.

If we keep the charge constant and alter the plate spacing, as shown at B, the potential will vary inversely with the change in capacitance. If we make one plate stationary and the other variable, as a diaphragm that can vibrate with sound pressure changes, we have the makings of a microphone, as shown at C.

The output power of a microphone is extremely small, so it is customary to amplify the signal at the capacitor element itself. Two forms of this are shown: at D, we show the conventional approach in which a *polarizing*, or bias, potential supply is required to maintain the charge. At E, we show the *electret* princple. Here, the capacitor is permanently charged (in a way analogous to a magnet) by the action of a special

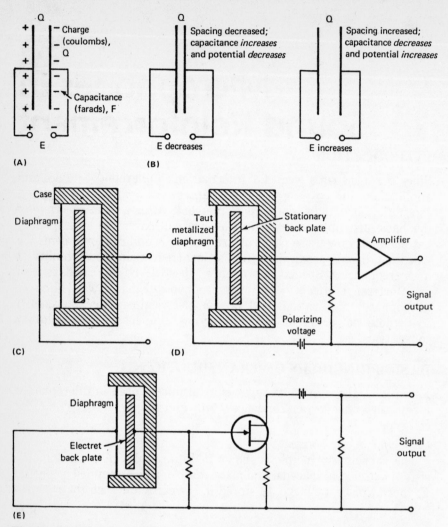

Figure 8-1. The principle of the capacitor microphone. **(A)** A charged capacitor. **(B)** Changing plate spacing (capacitance) changes potential. **(C)** A capacitor with a diaphragm and a fixed back plate. **(D)** A conventional capacitor microphone. **(E)** An electret capacitor microphone.

material applied to the back plate. Such a microphone requires only a small potential, perhaps no more than 6 to 8 volts, to operate the amplifier, while the conventional capacitor microphone will require up to 48 volts to polarize it.

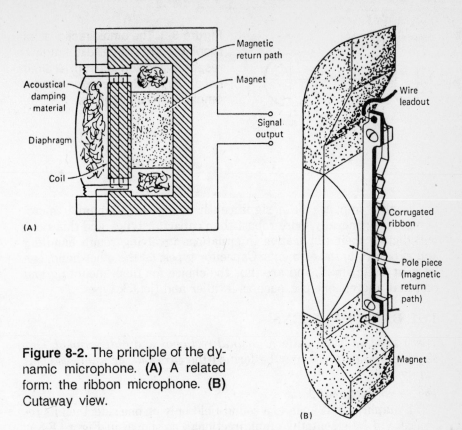

Figure 8-2. The principle of the dynamic microphone. **(A)** A related form: the ribbon microphone. **(B)** Cutaway view.

Dynamic microphones:

If a coil of wire is attached to a diaphragm, and if that coil is placed in a magnetic field, then, through the principle of magnetic induction, the moving diaphragm will produce an output potential across the coil equal to:

$$E = Blv, \tag{8-2}$$

where E is the output potential, B is the flux density of the magnetic circuit in teslas, l is the coil length in meters, and v is the velocity of the coil motion in meter/sec.

The principle of the moving coil, or dynamic, design is shown in Figure 8-2A, while a similar arrangement, the ribbon microphone is shown at B.

At one time, capacitor microphones were felt to be superior to dynamic types in overall bandwidth and quality. But in recent years

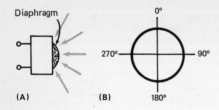

Figure 8-3. The omnidirectional microphone. **(A)** A small diaphragm is responsive to sounds from all directions. **(B)** A polar representation of omnidirectional response.

the quality gap has been significantly narrowed. Dynamic microphones are generally more robust than capacitor types, and this alone will dictate their application in situations involving rough handling and exposure to the elements. Capacitor types, on the other hand, can be kept quite small, and are thus the choice for flush mounting and personal microphone use, such as lavalier and tietack types.

THE BASIC PATTERNS

There are two basic patterns, *omnidirectional* and *bidirectional*. Out of these two patterns may be formed the *cardioid*, or *unidirectional* family.

The omnidirectional pattern:

If a diaphragm is open to a sound field only on one side, then its response will be essentially omnidirectional, as shown in Figure 8-3A. Typical models today have diaphragms on the order of 20mm in diameter or less. In polar coordinates, the omnidirectional pattern is as shown at B.

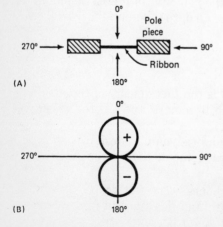

Figure 8-4. The bi-directional microphone. **(A)** Top view of a ribbon microphone. **(B)** Polar representation of bi-directional response.

The bidirectional pattern:

The ribbon microphone is probably the best known form of bidirectional microphone. Note in Figure 8-4A that the ribbon is open to the sound field on two sides. Obviously, sound waves approaching from 90 or 270 degrees will result in no output from the microphone, since pressures will cancel on both sides of the ribbon. The polar pattern of the bidirectional microphone is shown at B. Note the characteristic "figure-8" shape of the pattern.

The cardioid pattern:

As shown in Figure 8-5A, the cardioid, or unidirectional, pattern can be produced by adding the omnidirectional and bidirectional patterns together. While most variable pattern microphones operate on the principle of combining the two elements, the fixed pattern cardioids widely used in sound reinforcement today are of the single diaphragm type. Their principle of operation is shown in Figure 8-5B and C.

In this design, the back of the diaphragm is open to the sound field, but in such a way that the rear sound path to the diaphragm has an acoustical delay equal to that around the microphone from back to front. As a result, sound approaching from the back side arrives at the diaphragm at the same time via the two paths, and there is no net output. For sounds arriving from the front, there is, of course, some cancellation, but there is a net pressure available to drive the diaphragm. Typical 0, 90, and 180-degree response for a cardioid microphone is shown at D.

Summary of the first-order cardioid family:

By varying the delay path in a single-diaphragm cardioid microphone, we can arrive at the *hypercardioid* and *supercardioid* patterns, shown in Figure 8-6. The hypercardioid pattern exhibits the greatest ratio of on-axis pickup to total random response, while the supercardioid exhibits the maximum front hemisphere response to total random response.

The patterns we have discussed so far are known as *first-order* designs. They are so-named because they can be derived from the simple omnidirectional and Figure-8 patterns. Higher-order patterns would involve squaring the cosine term in the polar equations for the microphone patterns shown in Figure 8-6. Higher order patterns are more directional than the first-order patterns, but they are quite hard to design and exhibit their ideal patterns only over a small portion of the frequency range.

The chart shown in Figure 8-7 summarizes the first-order family. The normal cardioid pattern would be used where there are unwanted sound sources at 180 degrees to the principal source. The hypercardioid pattern would be especially useful in a highly reverberant environment,

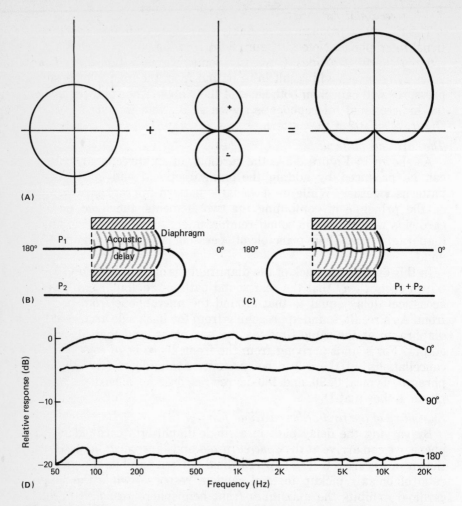

Figure 8-5. The uni-directional or cardioid microphone. **(A)** Omni pattern plus bi-directional pattern equals uni-directional pattern. **(B)** A single-diaphragm cardioid microphone; for sounds arriving at 180°, delay path P_1 is equal to outside path P_2, and signals cancel at the diaphragm. **(C)** For sound arriving at 0°, total rear delay is $P_1 + P_2$; there is thus some cancellation, but the net result is sufficient pressure to drive the diaphragm. **(D)** Typical response of a good cardioid microphone.

since it has the highest discrimination against random sounds. The supercardioid pattern is useful in picking up a wide-angle source, such as a chrous, while maintaining good rejection of reverberant information over its back hemisphere.

148

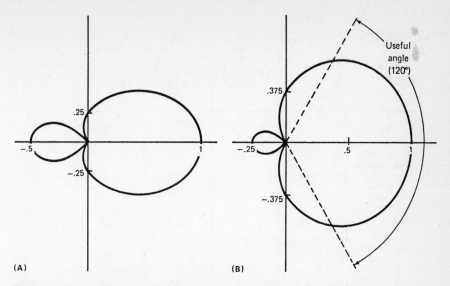

Figure 8-6. Hyper-cardioid and super-cardioid patterns. **(A)** The hyper-cardioid pattern. **(B)** The super-cardioid pattern.

RANDOM ENERGY EFFICIENCY (REE) AND DISTANCE FACTORS

These entries in the chart of Figure 8-7 may need clarification. REE is a measure of the ratio of on-axis pickup to total random pickup characteristic of a given pattern. It is analogous to the directivity factor of a loudspeaker, or to DI, if expressed in decibels. The Distance Factor is shown in Figure 8-8. Distance factors express the maximum working distances of directional patterns that will result in the *same* immunity to random response as compared to the omnidirectional pattern.

PROXIMITY EFFECT

When directional microphones are used at close working distances, there is a rise in their LF response. This is due to a complex interaction between pressure differences at the front and back of the diaphragm as influenced by close-in inverse square relationships. Figure 8-9A shows the theoretical curves for a cardioid pattern as a function of angle for a fixed operating distance. At B, the response is shown for a typical microphone intended for relative close sound reinforcement use. Note that the microphone has been designed to be rolled off at low frequencies for distant sources. At a distance of about 0.6 meter (2 ft.) the response is only slightly rolled off (about –5 dB at 100 Hz), indicating

149

CHARACTERISTIC	OMNI-DIRECTIONAL	CARDIOID	SUPER-CARDIOID	HYPER-CARDIOID	BIDIRECTIONAL
Polar response pattern					
Polar equation	1	$.5 + .5 \cos \theta$	$.375 + .625 \cos \theta$	$.25 + .75 \cos \theta$	$\cos \theta$
Pickup ARC 3 dB down (1)	—	131°	115°	105°	90°
Pickup ARC 6 dB down	—	180°	156°	141°	120°
Relative output at 90° dB	0	−6	−8.6	−12	−∞
Relative output at 180° dB	0	−∞	−11.7	−6	0
Angle at which output = 0	—	180°	126°	110°	90°
Random energy efficiency (REE)	1 0dB	.333 −4.8dB	.268 −5.7dB(2)	.250 −6.0dB(3)	.333 −4.8dB
Distance factor (DF)	1	1.7	1.9	2	1.7

NOTE:
1 = Drawn shaded on polar pattern
2 = Maximum front-to-total random energy efficiency for a first order cardioid
3 = Minimum random energy efficiency for a first order cardioid

Figure 8-7. The family of first-order cardioids and their components.

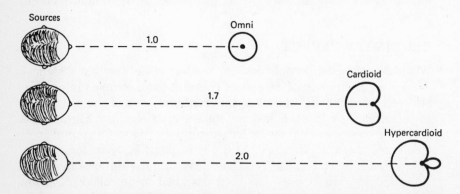

Figure 8-8. Illustration of distance factor.

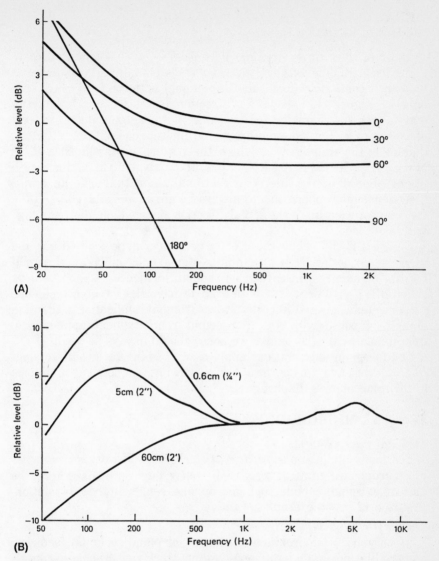

Figure 8-9. Proximity effect in cardioid microphones. **(A)** Calculated plots of proximity effect for a uni-directional microphone at 60 cm (2 ft). **(B)** Proximity effects in a typical cardioid microphone.

that this microphone would have fairly flat response if used at that distance in a speech reinforcement system.

The omnidirectional pattern exhibits no proximity effect.

DEPARTURES FROM IDEAL RESPONSE
AT HIGH FREQUENCIES

Because of their finite size, all microphones' patterns tend to become more directional on-axis at high frequencies. Omnidirectional patterns tend to become directional, and directional patterns tend to collapse on-axis. Figure 8-10 shows the directional pattern, as a function of wavelength and diameter, for a circular diaphragm located at the end of a long tube. For an omnidirectional microphone of 20mm (0.8 in.) diameter, the frequency at which the pattern has assumed a REE of 0.1 (DI = 10 dB) is 16.5 kHz. Thus, above about 10 kHz, this microphone should be considered directional. In actual use, an omnidirectional microphone should always be aimed at its target, and the user must always know the actual direction along which this "beaming" will take place.

Figure 8-11 shows how the effect of beaming can be dealt with by the microphone designer. If an omnidirectional microphone is designed for flat on-axis response, then its random incidence response will be rolled off. Conversely, if it is desired to have flat random incidence response (as is needed in certain measurement applications), then the on-axis response will rise. For sound reinforcement applications, microphones with flat on-axis response should always be used.

The foregoing observations apply as well to the various directional microphones, and the degree of HF narrowing is dependent on the size of the microphone's diaphragm.

SPECIAL MICROPHONE TYPES

Personal microphones:

For many applications, *lavalier, tietack,* or head-worn *swivel* types of microphones may be useful. In noisy environments where the user must remain mobile, such microphones represent the only solution. Figure 8-12 shows examples of these.

The noise-cancelling microphone:

In very noisy environments, such as airplane cockpits, factories, and the like, normal microphones will not work well. The noise-cancelling microphone, shown in Figure 8-13, handles most of the problem by discriminating against random LF noise in favor of close-in speech. As a field expedient, a pair of reversely-wired omnidirectional microphones will provide a reasonable degree of noise-cancelling action.

Figure 8-10. Directional properties of a circular diaphragm at the end of a tube as a function of diameter and wavelength (λ)

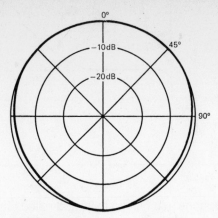

(A) Diameter = λ/6.3
REE = 0.77
DI = 1.1dB

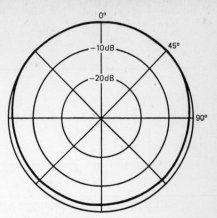

(B) Diameter = λ/3
REE = 0.53
DI = 2.8dB

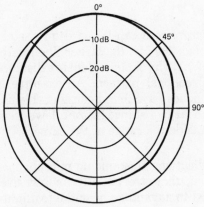

(C) Diameter = λ/2.0
REE = 0.33
DI = 4.8dB

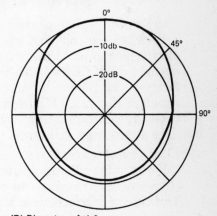

(D) Diameter = λ/1.6
REE = 0.22
DI = 6.6dB

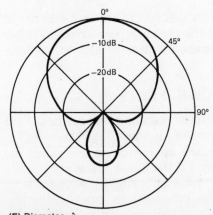

(E) Diameter = λ
REE = 0.1
DI = 10dB

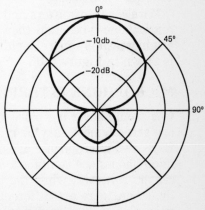

(F) Diameter = 1.2λ
REE = 0.06
DI = 12dB

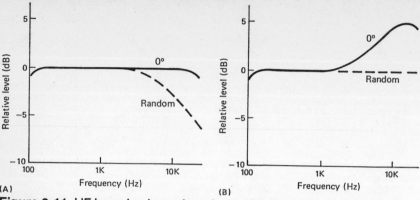

Figure 8-11. HF beaming in a microphone. **(A)** Flat on axis, and rolled off for random incidence or off-axis signals. **(B)** The response of a peaked microphone displays a rising on-axis response, and a flat response for random incidence or off-axis signals.

Distant pickup microphones:

So-called *line*, or "shot gun," microphones can be used for certain distant pickup problems. Above about 200 Hz, the larger line microphones exhibit REE's on the order of 0.1. This is equivalent to a distance factor of 3.2 relative to an omnidirectional pattern. An example is shown in Figure 8-14. Such microphones have found considerable use in television broadcasting, where the large distance factor allows them to be aimed from a distance well outside the range of the camera.

Flush-mount microphones:

For certain podium and altar applications, flush-mount microphones offer excellent response. In addition, they are visually unobtrusive. This type of microphone is also used to advantage near the footlights for both speech and music pickup in theater environments. An example is shown in Figure 8-15.

Wireless microphones:

Wireless transmitters and receivers can be used with a number of microphone types, and they afford the user complete freedom. They are increasingly more popular for stage use in musicals. An example is shown in Figure 8-16.

MICROPHONE INTERFERENCE EFFECTS

One should never use more microphones in sound reinforcement than are actually necessary. A typical problem is shown in Figure 8-17A. Here, two microphones are used to pick up a single talker. When the microphones are combined, all is good if the talker is *precisely* in the center. However, when the talker moves only slightly to one side or the other, there are different path lengths between him and the two microphones. The resulting response of the summed microphones is

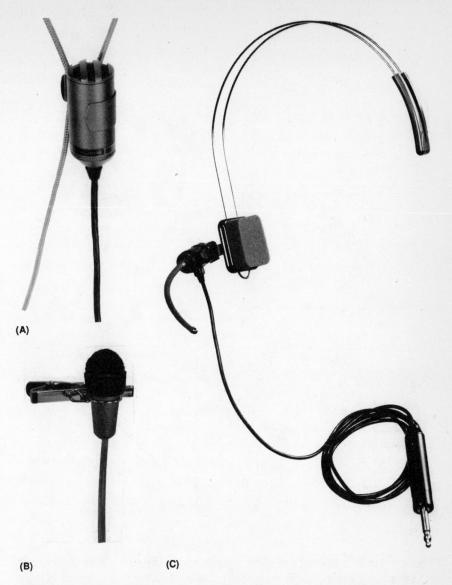

(A)

(B)   (C)

Figure 8-12. Personal microphones. **(A)** Lavalier style (Shure Bros. photo). **(B)** Tie-tack or lapel style (E-V photo). **(C)** Boom style (E-V photo).

shown at B. If an especially wide pickup angle is required, the microphones may be placed one atop the other and splayed, as shown at C. The close spacing will minimize path length differences, and the resulting response will be smooth, except possibly at the highest frequencies. Wherever possible, though, a single microphone should be used.

155

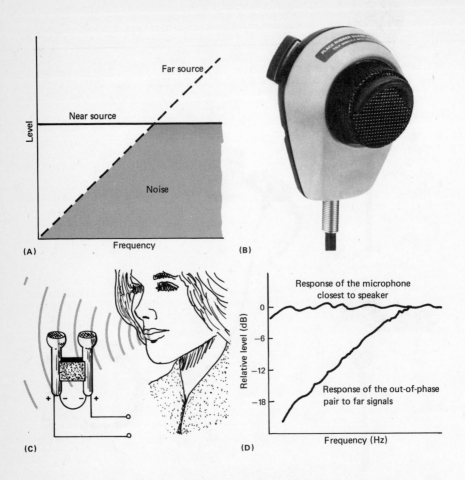

Figure 8-13. The noise-cancelling microphone. **(A)** The response to near and far sound sources. **(B)** A typical noise-cancelling microphone (Shure Bros. photo). **(C)** Out-of-phase wiring of two omni-directional microphones to produce a noise cancelling effect. **(D)** The effective response of the pair.

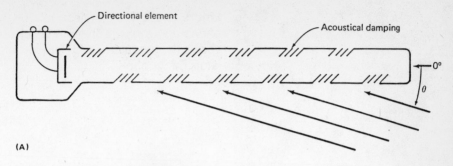

(A)

(B)

Figure 8-14. The line or "shot-gun" microphone. **(A)** Sounds entering at zero degrees reach the microphone element with less attenuation than sounds incident at off-axis angles. **(B)** The Electro-Voice CL42S microphone (E-V photo).

Figure 8-15. A flush-mount microphone (Crown photo).

157

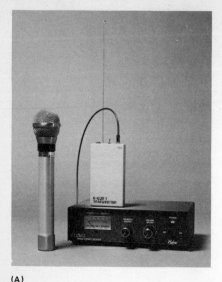

Figure 8-16. The wireless microphone and a block diagram which
shows the modulator and demodu-
lator (Edcor photo).

(A)

(B)

Another frequently encountered problem is the presence of a reflected, delayed signal entering a microphone along with the intended
direct sound. The usual problem is shown in Figure 8-18A.

Another example of the problem is shown at C.

The use of directional microphones is generally recommended where
reflections are likely to interfere with the main signal. However, for
podium use, a flush-mounted microphone will give the best results.

MICROPHONE SPECIFICATIONS

Microphone sensitivity ratings:

There is no single accepted method of specifying the sensitivity of a
microphone. Older rating methods give the microphone's power
output level when the microphone is placed in a specified sound field,
while newer rating methods state the microphone's output potential
when it is placed in a specified sound field.

Today, most microphone input circuits do not load the microphone
significantly; in fact, the average input circuit is high enough in

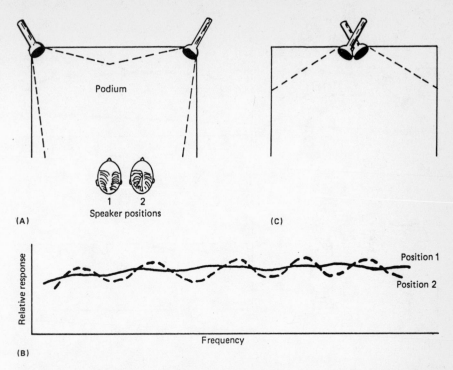

Figure 8-17. Multi-microphone interface effects. **(A)** Two cardioids improperly used at a podium. **(B)** Summation of microphones. **(C)** Cardioid microphones, one on top of the other, splayed.

impedance so that the microphone can be said to be operating into an open-circuit.

We will now consider an open-circuit sensitivity specification for a microphone:

Microphone sensitivity: .007 volts, 1 pascal

What this specification means is that the microphone will produce an output potential of .007, or 7 millivolts, when it is placed in a sound field of one pascal, or 94 dB SPL.

A typical open-circuit sensitivity rating for a dynamic microphone might read:

Microphone sensitivity: –80 dB (0 dB = 1 volt/microbar)

A sound field of one microbar is equivalent to a level of 74 dB SPL. The open-circuit potential will be:

antilog $(-80/20) = 10^{-4} = .0001$ volts

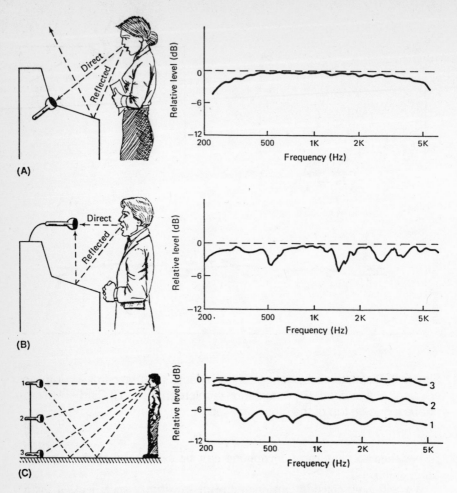

Figure 8-18. Interference by reflections. **(A)** Proper mounting of a podium microphone. **(B)** Raising the microphone creates a delay path and potential interference. **(C)** Floor stand microphone problems; as microphone height decreases, frequency response improves (response curves) as less and less reflected sound reaches the microphone.

If we want to compare this rating with the previous one, we first note that the previous rating was made in a sound field of 94 dB SPL, while the second was made at 74 dB SPL. This level difference of 20 dB must be corrected and this indicates that the 74 dB rating must be *raised* 20 dB, or multiplied by a factor of 10. When this is done, the second microphone's equivalent 94 dB rating will be .001 volts. Now we can compare the two. At .007 volts output, the first microphone is some 17 dB more sensitive than the second microphone:

20 log (.007/.001) = 17 dB

Some manufacturers give microphone sensitivity ratings by stating the power level in dBm the microphone can deliver to a matching load when it is placed in a reference sound field. Such a specification might read:

Impedance = 200 ohms
Power output = –40 dBm (10 dynes/cm²)

First, let us note that 10 dynes/cm² is equivalent to one pascal, or 94 dB SPL. The power level of –40 dBm indicates that the microphone output, when it is placed in a 94 dB sound field, will be 40 dB *below* one milliwatt. The power corresponding to this is:

antilog (–40/10) = $10^{\frac{-40}{10}}$ = .0001 milliwatt = .1 × 10⁻⁶ watt.

Since power = E^2/Z, we can solve for the output potential across the load of 200 ohms:

$$E^2 = (.1 \times 10^{-6}) \times 200 = 20 \times 10^{-6}$$

Therefore, E = .0045 volts.

This is the potential across the 200-ohm load; when the microphone is unloaded, the value will *double* to .009 volts.

Figure 8-19A presents a useful nomograph for determining both loaded and unloaded potentials for microphone power levels in a 94 dB SPL sound field as a function of microphone impedance. Simply place a straight edge across the chart intersecting the power level and the impedance. Then, at the left scale, read the voltage level. These voltage levels may be converted directly into volts or millivolts using the nomograph shown at B.

Microphone impedance ratings:
So-called low-impedance, or low-Z, microphones will have impedances ranging between 50 and 200 ohms. High impedance microphones are usually in the 10,000-ohm range, and such models are not recommended for use in sound reinforcement. A microphone may be safely loaded with its rated impedance, but most input circuits encountered today have an impedance of about 3000 to 3500 ohms. As stated earlier, this value is high enough so that, from the point of view of a low-impedance microphone, it is virtually an open circuit; therefore, a precise knowledge of the microphone's impedance is of little use. Only when we are interpreting microphone power ratings will an exact statement of the impedance rating be required.

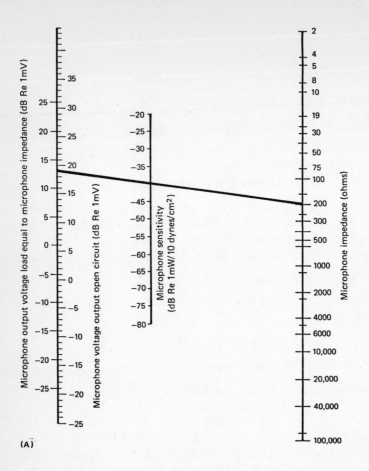

Figure 8-19. Nomograph for determining microphone output levels and potentials. **(A)** Relating microphone power output level with impedance and potential. **(B)** Relating microphone potential level with volts and millivolts.

Microphone noise ratings:

Noise ratings are not normally stated for dynamic microphones, since they are not usually specified for critical applications, such as recording of low-level sounds. Capacitor microphones intended for recording use will have noise specifications indicating their self-noise level as an equivalent acoustical noise level as measured on the A-weighting scale of a sound level meter. Such a rating may be referred to as a dB(A) rating.

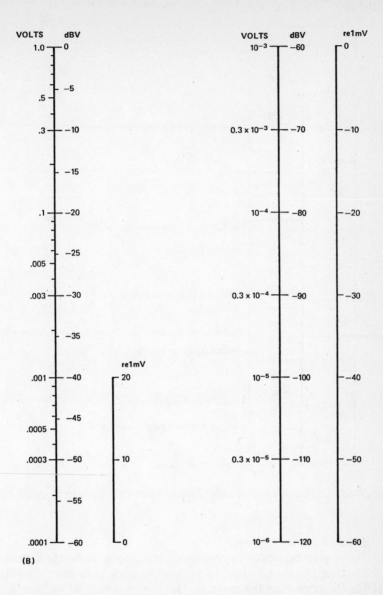

(B)

For example, a high-quality capacitor microphone might have a noise rating of 17 dB(A). What this means is that the self-noise of the microphone (that noise inherent in its preamplifier input stage) is equivalent to placing a noise-free microphone into an acoustical sound field of 17 dB(A). Typical noise ratings for both conventional capacitor microphones and elecret microphones will be in the 17 to 27 dB(A) range. The nomograph of Figure 8-20 gives some idea of the acoustical levels of a variety of sound sources as they relate to microphone characteristics.

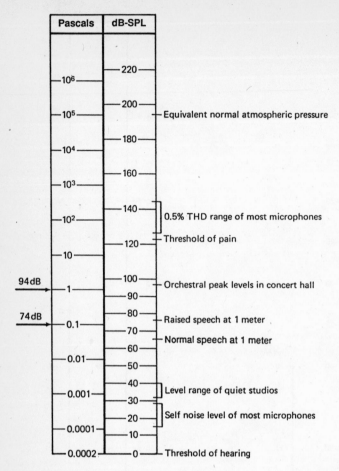

Figure 8-20. Levels of various sound sources as the relate to microphones.

Microphone distortion ratings:

Studio quality capacitor microphones are usually rated for distortion by indicating the sound field in which the total harmonic distortion (THD) is no more than 0.5 percent. Usually, this will take place in the 130 to 140 dB SPL range.

Most capacitor microphones have an internal pad which can be inserted to increase the allowable operating level by 10 to 15 dB. However, when the pad is used, the noise floor is raised by an equivalent amount.

Many dynamic microphones can operate in extremely high sound fields with little apparent distortion, since they tend to overload rather gently as compared to capacitors.

Examining the overall dynamic range of microphones, a high-quality capacitor microphone intended for studio use will have a total range up to 110 dB, and this is far greater than any studio recorder in use today—and certainly far greater than will ever be required in sound reinforcement.

REMOTE POWERING OF CAPACITOR MICROPHONES

Most speech and music input consoles in use today have provision for 48-volt remote powering for capacitor microphones. The basic circuit for this is shown in Figure 8-21. The 6800-ohm resistors are matched within 1 percent, and with a low-impedance DC power supply, crosstalk between inputs can be held to quite low values. Since there is no net potential difference between pins 2 and 3 (the signal-carrying pins), dynamic microphones may be safely plugged into a remote powering circuit with no adverse effects.

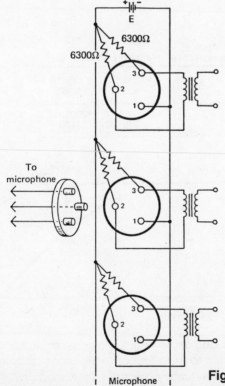

Figure 8-21. Remote powering of capacitor microphones.

165

LINE LOSSES

If high-quality microphone cable is used, line losses should be negligible in any sound reinforcement system. Typical good quality microphone cable will be made up of a pair of 24-gauge (AWG) stranded copper wire placed in a braided shield. The resistance per-unit length of the inner conductors will be about .08 ohms per-meter, and the capacitance between the conductors will be about 100 picofarads per-meter. In the equivalent circuit of Figure 8-22A, note that we have ignored the effects of line resistance, since the values are quite small compared with resistance of the load and the source impedance of the microphone.

At 1 kHz, the effect of the 1,000 picofarad capacitance between 10-meter conductors is negligible; it amounts to a reactance of $-j160,000$ ohms across the resistive load of 3,000 ohms. At 20 kHz, the capacitive reactance will be $-j8,000$ ohms across the resistive load of 3,000 ohms.

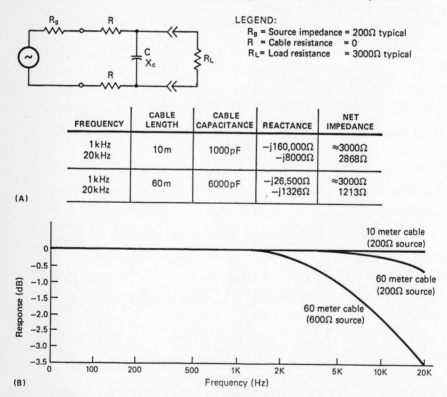

LEGEND:
R_g = Source impedance = 200Ω typical
R   = Cable resistance = 0
R_L= Load resistance   = 3000Ω typical

FREQUENCY	CABLE LENGTH	CABLE CAPACITANCE	REACTANCE	NET IMPEDANCE
1 kHz 20 kHz	10 m	1000 pF	$-j160,000Ω$ $-j8000Ω$	≈3000Ω 2868Ω
1 kHz 20 kHz	60 m	6000 pF	$-j26,500Ω$ $-j1326Ω$	≈3000Ω 1213Ω

(A)

(B)

Figure 8-22. Line losses. **(A)** In a long cable length, the cable capacitance high frequencies are attenuated due to the effect of cable capacitance on the system's net impedance. **(B)** Frequency response for various cables and source impedances.

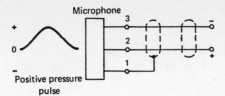

Microphone

Positive pressure pulse

Figure 8-23. Microphone polarity; a positive pressure pulse should produce a positive voltage at the terminal.

The net impedance of this combination will be about 2,870 ohms, as compared with the nominal value of 3,000 ohms. This is still large enough to be negligible.

Now we will consider the losses in a 60-meter (200 ft.) cable. Again, the line resistance is quite small compared with the 3,000-ohm load, and we will ignore it. At 1 kHz, the reactance of the 6,000 picofarad capacitance is –j26,500 ohms across the load resistance of 3,000 ohms, and the effect is negligible. However, at 20 kHz, the reactance will be –j1,326 across the resistance of 3,000 ohms, producing a net value of 1,213 ohms. The effect of this is a reduction of the signal at the load by about 1.3 dB.

The reason why the rather wide swing in load impedance has such little effect on the voltage at the load is that the source impedance is much lower than the load impedance, in this case by a factor greater than 10-to-1. Consider now the effect if our microphone impedance were 600 ohms instead of 200 ohms. At 20 kHz, the loss over 60 meters of cable would be a little in excess of 3 dB.

For very long cable runs, as in large theater complexes or large outdoor music reinforcement applications, many system designers will provide amplification for all microphones as close to the source as possible, with subsequent runs made at line level. Under these conditions, there will be virtually no loss over the longest runs.

MICROPHONE POLARITY CONVENTIONS

All the microphones used in a given sound reinforcement job should be identically poled. What this means is that a positive-going acoustical signal at the microphone should result in a positive-going potential at the *same* output pin on all microphones. The standard today is as shown in Figure 8-23. Pin 1 of the standard 3-pin connector is ground, and pins 2 and 3 carry the signal. Pin 2 should be positive-going when a positive pressure exists at the microphone. While most microphones come wired this way, it is always a good precaution to compare them with ones known to be correctly poled.

CHAPTER 8:

Suggested Reading:

1. G. Bore, *Microphones*, George Neumann GMBH, Berlin (1978).
2. J. Eargle, *Microphone Handbook*, Elar Publishing, Plainview, NY (1982).
3. Various, *Microphones*, (compiled from the pages of the Journal of the Audio Engineering Society, New York, 1979).

CHAPTER 9

System Architecture

INTRODUCTION

System architecture consists of the rational specification and layout of electrical and electroacoustical components to perform specific functions with minimum distortion and noise. The analogy with building architecture is a clear one, and as system architects we should be no less concerned with the user's needs and his budget.

To some extent, we have already introduced our subject in Chapter 7, where we discussed the matching of HF and LF elements through passive networks or through biamplification. In this chapter, we will extend the discussion to include microphone input and power output stages, noting carefully all operating levels of the active components.

Most of the system drawings in this chapter will be in the form of block diagrams, more appropriately known as signal flow diagrams. These drawings are fairly simple and are the best way to show relative levels in a system. Where circuit details are important, full schematic drawings will be used. These of course can be used in actual wiring and system hook-up procedures.

Further topics to be discussed include: metering considerations, interfacing of various signal processing devices, connectors and patch bays, shielding and grounding practice, and a short discussion of priority systems and remote control circuits.

DIVISION OF GAINS AND LOSSES

In any chain of active devices, such as preamplifiers, mixers, signal processors, and power amplifiers, there is more gain available than can be used. This reserve gain, or gain overlap, as it is often called, provides for flexibility in combining the active devices in various ways, and it makes possible the insertion of a variety of passive elements into the chain.

A Complex Audio Transmission System:

The proper division of gains and losses is at the very heart of good system architecture, and the three examples shown in Figure 9-1 will be instructive. We have shown a flow diagram for a single microphone channel through a complete recording system. The relative complexity of these systems draws special attention to the operational problems

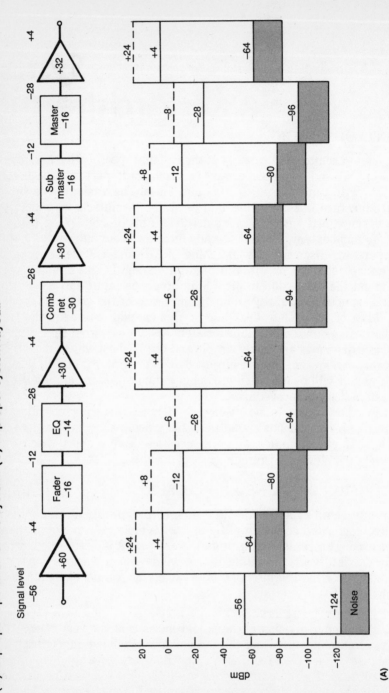

Figure 9-1. Gains and losses in a recording system. **(A)** Proper operation of the system. **(B)** Improper operation of the system. **(C)** Improper system layout.

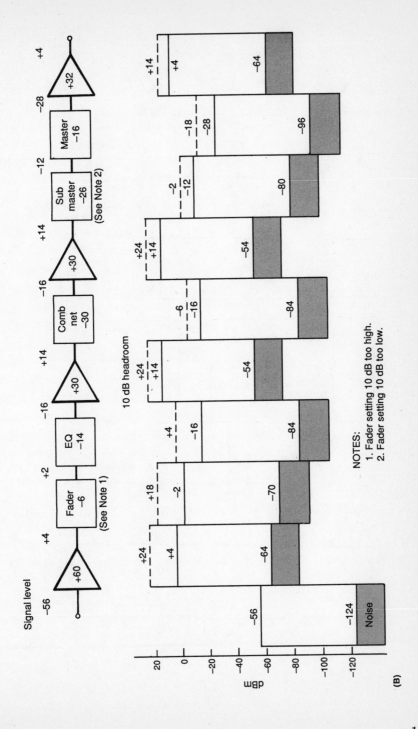

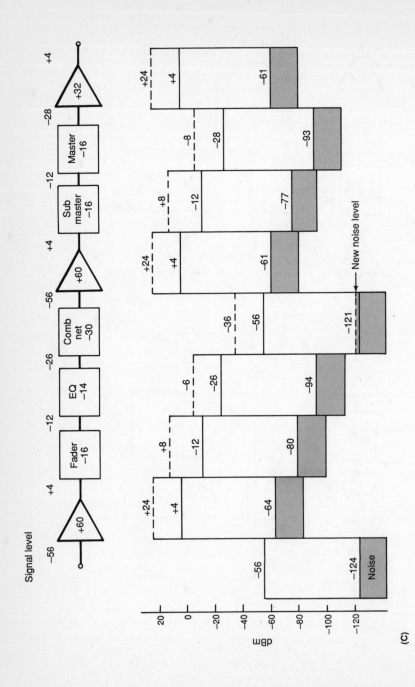

encountered when the system is improperly designed or improperly used.

The system shown at A is properly designed, and the fader settings are at their nominal operating positions. The signal levels in the diagram are in dBm, referred to 600 ohms. In this system, there is 20 dB "headroom" between the normal operating level and the clipping output level of any amplifier. Note that nominal settings of all faders are such as to provide that safety factor at all times, allowing for musical variations in level and consequent balancing requirements.

At B, the system is improperly operated. Note that the microphone fader is set 10 dB too high, while the sub-master fader is set 10 dB too low. What this means is that average levels are too high going into the booster amplifier which follows the equalizer, thus enabling it to overload too easily. The result is a 10-dB reduction in overall dynamic range of the system. This is a common mistake made by inexperienced operators.

At C, we have changed the system architecture slightly, replacing two 30-dB booster amplifiers with a single 60-dB booster amplifier. In our attempt to be more economical, we have allowed the noise level to drop down to a point equivalent to the input noise level; thus, we are effectively doubling our noise level, raising it some 3 dB.

The lessons taught by these three examples are very important, and the reader should understand them well before proceeding with this chapter.

A Simple Sound Reinforcement System

A simple sound reinforcement system might consists only of a multi-input microphone mixer, a power amplifier, and a loudspeaker, as shown in Figure 9-2. The signal-to-noise ratio should always be established at the input, and proper layout of the system from that point onward should follow these basic design rules, that:

1. the signal level be kept well above subsequent noise levels in the system,

2. peak signal level requirements have sufficient reserve, or headroom, in all active components before the onset of clipping, and

3. signal level metering be properly set so that it relates usefully to the amplified program level in the listening area.

In the system shown in Figure 9-2, we are using a dynamic microphone which has a rated power sensitivity of –53 dBm *re* 1 pascal. If we assume peak speech levels at the microphone of, say, 78 dB SPL, then the microphone's output level will be –53–(94-78), or –69 dBm. Since the mixer's input stage has a rated noise floor, or *equivalent input noise* (EIN), of –124 dBm, this indicates that the noise level will be some 55 dB below speech peaks. We will further assume that average speech levels will run about 10 to 12 dB lower than peak levels.

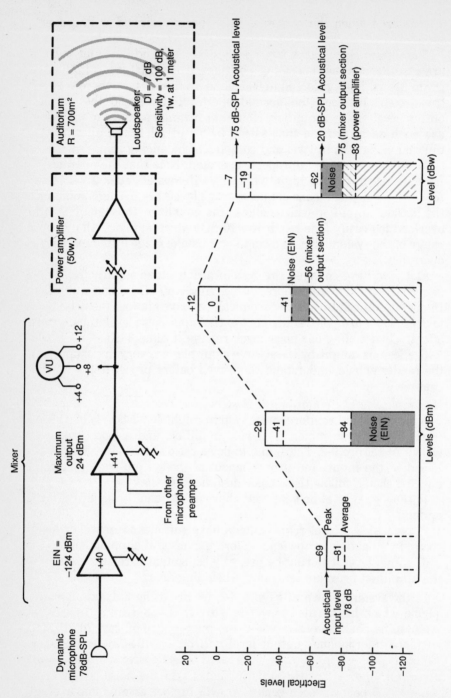

Figure 9-2. A simple sound reinforcement system.

In order to maintain the signal-to-noise ratio established at the input, we should endeavor to operate all subsequent active devices at a high level so that the noise floor does not deteriorate. The mixer used in this example has a maximum output capability of 24 dBm. In order to be on the safe side, we should set the mixer's output gain control so that peak speech levels produce an output of 12 dBm. When this is done, average speech levels will be in the range of 0 dBm. By setting gains in this manner, we are allowing 12 dB of reserve, or headroom, in output capability over anticipated speech peaks. At the same time, we are keeping the signal well above subsequent noise levels.

Note that the VU meter sensitivity switch has positions of +4, +8, and +12 dBm for readings of zero on the VU meter. Setting this switch to the +12-dB position would be a convenient starting point for final system calibration, inasmuch as speech peak levels would then produce a convenient zero reading on the VU meter.

The gain through the mixer is set at 81 dB, and it is divided between two sections, the input preamplifier and the output section. Normally, the user cannot measure the level at the output of the preamplifier, and he will usually adjust the input pads or trimmers on the pre-amplifier so that the level control, if of the rotary type, is operated in the "10 to 2 o'clock" range, with the rest of the required gain being set by the master output control. If it can be measured, or observed, as in the case of a linear straight-line fader, the optimum operating position for the preamp fader will be about 10 dB down from its maximum position. This will allow maximum flexibility in dealing with a variety of input speech levels.

There is, of course, additional noise introduced in the mixer's output stage and in the power amplifier. The specifications for most mixers used in sound reinforcement will state the noise present at the output with all of the microphone input controls set to zero. In the case we are working here, let us assume that this specification read: "noise 80 dB below full output; input controls set at minimum and output control set at maximum." While the output control may not be set at its maximum position, let us assume that it is; thus we have our worst case noise contribution from the mixer's output section. Since the maximum output from the mixer is +24 dBm, the noise level will be 80 dB lower, or −56 dBm. Note that this is 13 dB *lower* than the EIN of −43, and is therefore negligible.

The power amplifier's noise specification reads: Noise 100 dB below full output. In dBW, the amplifier's output level is +17. 100 dB below this level is −83 dBW. Again, this is well below the EIN as it is presented at the power amplifier's output; it is clear that the dominant noise present at the system's output is due to the noise level established at the system's input.

The final step in system calibration is relating the zero reading on the VU meter to a reference acoustical level in the audience area. We note that the power amplifier in this system has a potentiometer at its input; thus, there is no danger of overloading its input stage with the signal from the mixer. Let us assume that our design goals call for peak reverberant speech levels in the auditorium of 75 dB SPL. We have thus determined that a mixer output level of 12 dBm relates directly to an acoustical level in the audience area of 75 dB SPL. The corresponding input level produced by the talker at the microphone remains 78 dB SPL.

Taking into account the loudspeaker's sensitivity and directivity characteristics, as well as the acoustical characteristics of the listening space, we can complete the example by determining the power amplifier output level that will produce peaks at 75 dB SPL in the audience area, using the following equation:

$$dBW = SPL_r - 17 - S + DI + 10 \log R \qquad (9-1)$$

In this equation, SPL_r is the desired reverberant level in the room, S is the sensitivity of the loudspeaker, one watt at one meter, DI is the directivity index of the loudspeaker, and R is the room constant in square meters.

In English units this equation is:

$$dBW = SPL_r - 26 - S + DI + 10 \log R \qquad (9-2)$$

where S is the sensitivity referred to one watt at four feet, and R is the room constant in square feet.

Solving our example yields a power level of -7 dBW. This level in dBW can be related directly to power in watts by the equation:

$$\text{Power (watts)} = 10^{\frac{dBW}{10}} \qquad (9-3)$$

Thus, in this example, only 0.2 watt will be required in routine use of the system. However, in normal specification of sound reinforcement componentry, we would not want to choose an amplifier smaller than, say, 50 watts, since system requirements are often changed with time. Remember that we had earlier allowed excess output capability of 12 dB at the mixer's output in anticipation of changing requirements.

A More Complex Example with Biamplification:

Figure 9-3 shows details of a more complex system. The design goals are the same as in the previous example, the difference being the inclusion of an active frequency dividing network between the mixer and the power amplifiers. In this case, the mixer output level must be carefully controlled so that it never exceeds the maximum input level capability of the active dividing network. In this example, the maximum input level the electronic dividing network can accept is 20 dBm;

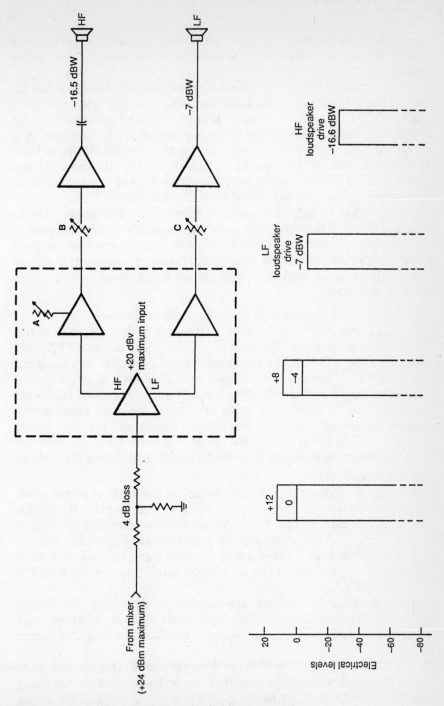

Figure 9-3. Adding an electronic dividing network.

therefore, it is desirable (but not absolutely essential) to construct a 4-dB loss pad (see Chapter 1) and locate it between the two active devices.

The LF path through the active network has unity gain, so the LF part of the system will be calibrated as before, taking into account the 4-dB pad. The HF path has a gain control which in effect is in series with the input potentiometer of the HF power amplifier. A combination of these two controls should be used to set the appropriate HF drive level. In this case, the HF loudspeaker sensitivity of 109 dB, one watt, one meter, and DI = 7 dB, would require a drive level of some –16.5 dBW, or about 0.02 watts, as calculated from Equation 9-3.

Among other things, this example shows how little power may be required most of the time by simple speech reinforcement systems in reasonably small spaces. The fact that most systems are over-designed in this respect is probably a good thing, taking into account the great variability of input conditions and the occasional need to restructure gain requirements.

A More Complex System with Broadband Equalization:

Thus far, we have used active components whose response is flat across their passbands. The uses of system equalization will be discussed in Chapter 11, but for the present we will consider only electrical consequences of the use of equalization. Figure 9-4A shows a typical broadband equalization setting as it might apply to a sound reinforcement system. In calculating gains and losses, we must consider the maximum levels as they pass through the device, and make all judgements based on these worst case levels. Figure 9-4B shows how the equalization scheme shown in A would be integrated into the system.

dBm, dBv, and dBV:

In many level diagrams, system designers make use of power levels in dBm as well as in their corresponding voltage levels, dBv. In this latter case, it is important to remember that the reference impedance is 600 ohms, whether or not a given active device is provided with an actual 600-ohm load. Figure 1-9 in Chapter 1 gives a scale of voltage levels in dBv. Note that 0 dBv is referred to 0.775 volt, while 0 dBV is referred to 1 volt.

Many professional devices are capable of delivering the output potential corresponding to fairly high levels of dBm; however, they cannot deliver the current necessary to produce the indicated power level.

A mixer may be specified as follows: Rated Output, 6.2 volts (+18 dBv). Such a rating implies that the device in question can easily deliver 6.2 volts—but not necessarily into a 600-ohm load. The device may be designed to be loaded with a bridging, or high impedance

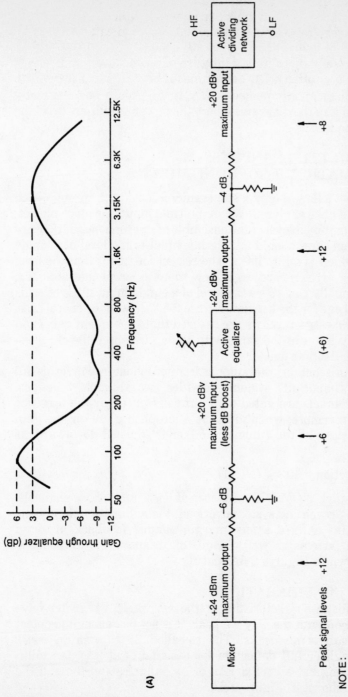

(A)

(B)

NOTE:

Pads ensure that no device can overdrive the next one in the chain. The first pad should be adjusted additionally to compensate for both gain through the equalizer and for the maximum output capability of the equalizer. When this is done, note that +24 dBm output from the mixer will not cause input or output overload in the chain.

Figure 9-4. Adding broadband equalization. **(A)** A typical eq curve. **(B)** Typical gain structure.

load, one that does not draw significant current. Manufacturers will often indicate the minimum impedance across which a given active device can deliver its rated output potential. Such indications should be carefully noted in laying out a system.

Level diagrams may contain a mixture of power levels and voltage levels, and the user often has to determine for himself which are which. When a level diagram is indicated in dBv, it will usually be consistently so up to the point in the diagrams where the power amplifiers drive the loudspeaker loads.

SUMMING MULTIPLE INPUTS:
THE AUTOMATIC MICROPHONE MIXER

So far, we have followed only a single microphone path through each system that we have analyzed. When multiple inputs, or microphones, are used, the combined level of the ensemble of microphones, if they are all on at a given time, should be roughly equal to the level of a single microphone which is on by itself. The reason for this is that the level in the listening area, regardless of how many microphones are open, should not normally rise above the level of a single input if the effect is to be a natural one for the listener. Figure 9-5A shows the general case. By following this structure, we are assured that subsequent gains and losses in the system are the same with the entire ensemble of microphones as with a single one.

Extending this notion, the automatic microphone mixer, a useful adjunct in a "hands-off" system, provides two valuable functions. Microphone channels are gated on or off as required by the presence of signal. Furthermore, when multiple microphones are on, the gain in the output stage of the automatic mixer is adjusted downward, as follows:

$$\text{Gain reduction} = 10 \log (\text{NOM}) \tag{9-4}$$

In this equation, NOM is the *number* of *open microphone* channels. In addition to preventing system overload, the automatic microphone mixer maintains a fixed safety margin against feedback in sound reinforcement systems as well. Details of an automatic microphone mixer are shown in Figures 9-5A and B.

METERING CONSIDERATIONS

While the ballistic characteristics of the venerable VU meter have much in common with the way we hear, it is not fast enough to detect potentially troublesome peaks. The so-called *peak program meter* (PPM), will register full deflection for a signal of at least ten milliseconds duration, and a comparison of its ballistics and those of the VU meter are shown in Figure 9-6A.

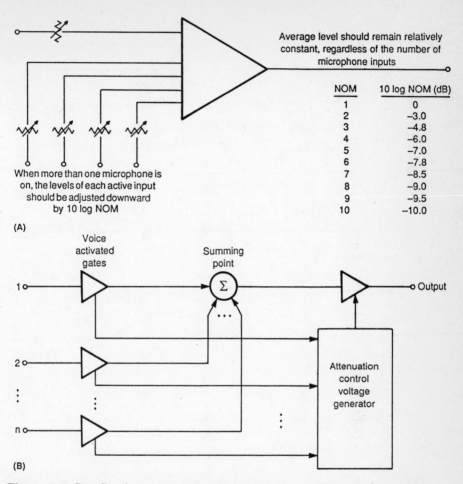

Average level should remain relatively constant, regardless of the number of microphone inputs

NOM	10 log NOM (dB)
1	0
2	−3.0
3	−4.8
4	−6.0
5	−7.0
6	−7.8
7	−8.5
8	−9.0
9	−9.5
10	−10.0

When more than one microphone is on, the levels of each active input should be adjusted downward by 10 log NOM

(A)

Voice activated gates

Summing point

1

Output

2

n

Attenuation control voltage generator

(B)

Figure 9-5. Details of an automatic microphone mixer. **(A)** Summing multiple inputs. **(B)** The principle of an automatic microphone mixer.

In instances where program peaks are not sudden or extreme, metering will not be a problem. This is the case indicated in Figure 9-6B, where the long-term program level proceeds by relatively small changes.

Occasionally, we encounter waveforms such as those shown at C; they are the trickiest to deal with since they will not be accurately indicated by peak program meters. The waveform shown is that of a trumpet playing a 400-Hz tone. The spread between the peak and average values of the waveform is about 13 dB, and such waveforms are a good argument for electrical headroom on the order of 20 dB.

181

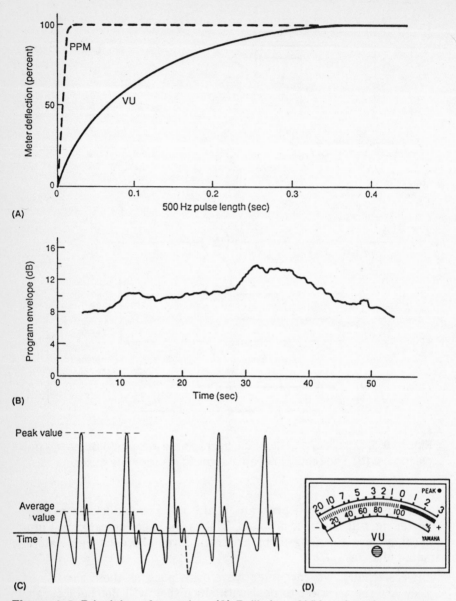

Figure 9-6. Principles of metering. **(A)** Ballistics of VU and peak-program meters. **(B)** An example of a slowly varying program envelope. **(C)** Steady-state waveforms with a peak factor of abut 13 dB. **(D)** A combination VU-peak indicator (courtesy Yamaha).

Similar problems are routinely encountered in music reinforcement, and for this reason we see more reinforcement consoles, as well as smaller mixers, which incorporate both VU and peak metering. Many times, the peak indicator is simply a single LED (light-emitting diode) indicator on the face of the VU meter set to blink at some peak level above zero on the VU meter, as shown at D. While not absolutely essential for most speech reinforcement work, the flexibility of this kind of metering is usually well worth its cost.

A TYPICAL MIXING CONSOLE
FOR SOUND REINFORCEMENT

Many of the problems we have discussed thus far in this chapter are conveniently solved for the user who chooses a small console. Many of these devices are aimed generally at recording and small-scale post-production work; consequently, they may have some features that are not really appropriate for sound reinforcement work. Figure 9-7 shows the functional diagram for a small Yamaha console, which is typical of what is available from many sources.

Let us assume that this small console is to be used in a house of worship. Some of the functions which can be addressed are:

1. Choral and instrumental music can be recorded in stereo *via* the left and right output buses. A reverberation device may be used for recording, being fed a mix through the ECHO OUT bus and returned to the left and right program buses through one or possibly two EFFECTS IN jacks.

2. For the main speech reinforcement part of the system, the fold-back (FB) mixing bus could conveniently be used. (Normally, foldback is used to allow musicians to hear themselves on stage, but that function would not be required here.)

3. During the recording of musical performances, soloists, for example, can be slightly reinforced, as required, by mixing a small amount of the soloist's microphone into the FB bus.

4. For purposes of sound reinforcement, the feed to the FB mixing bus might be made in the pre-equalizer position. This would ensure that the in-line equalizers could not inadvertantly be used in the reinforcement function, but only for the recording function.

5. The SUB IN options would not be used in an installation of this size.

6. Note that input level control involves attenuation as well as feedback gain control around the input amplifier. This will ensure that microphone overload of the console's input stages will be minimized.

We have shown the console with power amplifiers and external signal

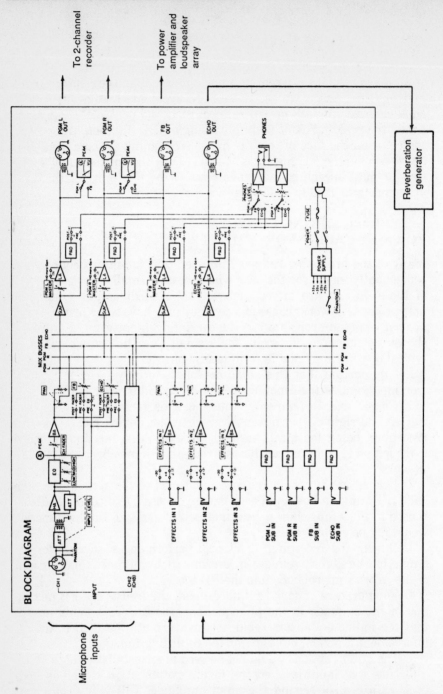

Figure 9-7. A small console for recording and sound reinforcement (courtesy Yamaha).

processors that would normally be used in a system of this size. The calibration procedure is exactly the same as in the previous examples.

If the above example is understood well, then it is quite easy to move on to larger, more complex, designs, inasmuch as the basic system architectural rules remain the same.

INTERFACING SIGNAL PROCESSING DEVICES

Depending on the application, a variety of signal processing devices may be used in conjunction with a basic sound reinforcement system. In addition to electronic dividing networks and equalizers, which we discussed earlier, such signal processors as compressors, limiters, reverberation generators, time delay units, and automatic gain control devices may be required for special purposes. Most of these ancillary devices operate at the console output at normal line output level (0 to +10 dBm).

Compressors and Limiters:

While there may be little need for compressors in speech reinforcement systems, they can be very useful adjuncts in certain music reinforcement applications. It is not uncommon, for example, for vocal soloists using hand-held microphones to require up to 10 or 12 dB of compression in order to maintain normal musical balances. Figure 9-8A shows the input/output characteristics of compressors and limiters. Note that a compressor gradually reduces output (with rising input), while the limiter provides a ceiling over which the output cannot rise.

Figure 9-8B shows the location of a compressor in a typical reinforcement system.

The dynamic characteristics of compressors and limiters are very important in determining their suitability for certain tasks in sound reinforcement. Aspects of attack and release time determine how natural the gain reduction function will be. For example, fast attack times, perhaps on the order of tens of milliseconds, are appropriate for a compressor, while the recovery time, when the input signal drops below the threshold of compression, may sound best when it is on the order of two or three seconds. In the case of a limiter, both attack and release times should be much faster than in a compressor, since it is usually the function of the limiter to prevent momentary overload.

Finally, the nature of signal level detection is important. RMS detection of signal level corresponds generally to the loudness of the signal and is well suited to use in compressors. Peak signal level detection is appropriate for limiters, since that mode of detection better fits the function of the limiter. Some designs incorporate both kinds of detection.

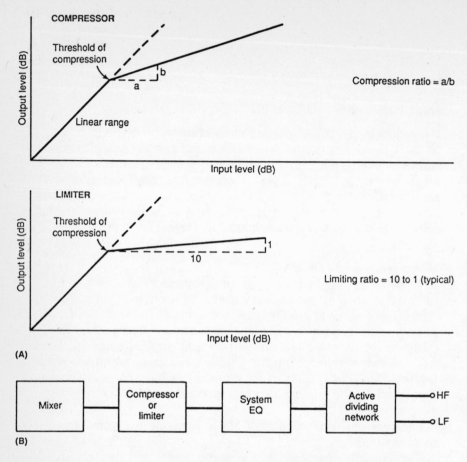

Figure 9-8. Compressors and limiters. **(A)** Input/output characteristics of compressors and limiters. **(B)** The location of a compressor or limiter in the audio chain.

Reverberation Generators:

Reverberation generators are rarely used on speech signals in sound reinforcement. However, they can be used to good advantage on certain kinds of reinforced music. The device is always placed in an external loop and fed back into the main mixing buses, as shown in Figure 9-9. Reverberation is normally a stereophonic effect, even when it is generated from a single channel signal. Care should be taken that the reverberation return signal cannot recirculate on itself, or be fed back into the reverberation device—unless, of course, that is the specific musical intention.

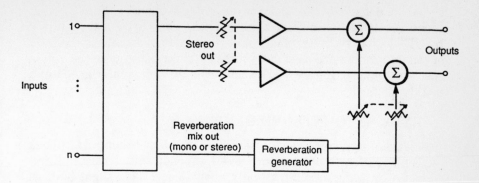

Figure 9-9. Interfacing a reverberation generator with an audio control system.

Time Delay Devices:

Delayed signals are fed to secondary clusters or arrays of loudspeakers in order to keep the listener's focus on sound arriving from the main array. For some applications, relatively inexpensive "bucket brigade" units can be used. However, where there is concern for the highest quality, digital models should be specified. Figure 9-10 shows a typical application of this device.

Automatic Gain Control Devices:

For paging applications in airports and other large terminals, the level of the paging signal must rise when the ambient noise level in the space rises. Logically, it should return to a comfortable listening level when the noise level in the terminal subsides. There are a number of different approaches which might be taken here. Perhaps the simplest method is to sample the noise level in the space and use that sampled

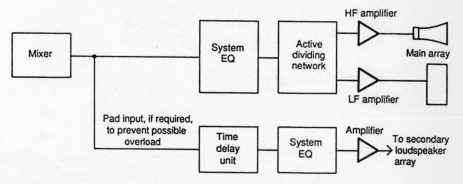

Figure 9-10. Applications of time delay.

signal to determine the gain of the paging channel. Figure 9-11A shows details of an ambient noise-controlled amplifier manufactured by UREI. Note the multiplicity of controls and adjustment on the front of the unit. These are used to trim the dynamic behavior of the unit to special situations. Figure 9-11B shows how the unit is interfaced into a paging channel.

RARELY ENCOUNTERED DEVICES

In an earlier day, frequency shifters were used to help control feedback in sound reinforcement systems. The frequency shifter raises all frequencies by some amount, usually about 5 Hz. This slight shift in frequency is barely perceptible on speech, but its effect on music may be an unpleasant one. Since feedback in an electroacoustical system depends upon specific phase and amplitude relationship in the loop, the function of the frequency shifter is to prevent the required zero phase angle from establishing itself. A system with a frequency shifter cannot necessarily be operated at a higher level than a system without one; it simply will not go into feedback as readily as one without the device.

For the same reason, Altec introduced a device which detects the presence of a feedback mode, or a sustained oscillation, and lowers the loop gain until the oscillation stops. The device then raises the gain gradually, maintaining a gain margin against feedback.

Devices such as these are not routinely specified in sound system design, but they may be very appropriate for special cases.

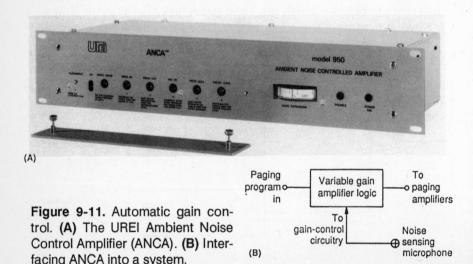

(A)

(B)

Paging program in → Variable gain amplifier logic → To paging amplifiers

To gain-control circuitry ← → Noise sensing microphone

Figure 9-11. Automatic gain control. **(A)** The UREI Ambient Noise Control Amplifier (ANCA). **(B)** Interfacing ANCA into a system.

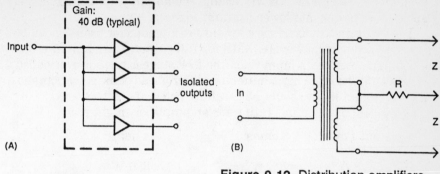

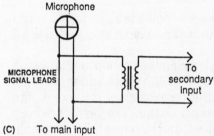

Figure 9-12. Distribution amplifiers and transformers. **(A)** A distribution amplifier. **(B)** A "hybrid" coil transformer; when R is properly adjusted, the two Z outputs remain isolated from each other; there is a 6 dB level drop through the transformer. **(C)** A signal-splitting transformer.

DISTRIBUTION AMPLIFIERS AND TRANSFORMERS

In many large applications, stage microphones may feed the reinforcement system, a recording system, and a broadcast system. In such cases, it is probably best to amplify all microphones up to line level and provide separate feeds to each of the systems. Distribution amplifiers are the best way to accomplish this. Typical models have four or more outputs, all independent of each other, as shown in Figure 9-12A. Distribution amplifiers have the further advantage of restricting long runs of microphone cables and raising the signal above the possible interference levels of lighting and other heavy power requirements.

To a more limited extent, certain transformers can provide two mutually isolated outputs. The price paid for this is a drop in level through the transformer. Details are shown in Figure 9-12B and C.

POWER AMPLIFIERS

General Characteristics:

By far, most of the power amplifiers encountered today are solid state designs, and their output power ratings are based upon their ability to swing a given potential across a stated minimum load impedance for a given amount of distortion. Most amplifiers will carry several power ratings, one for each nominal load impedance. Occasionally, an amplifier will have an output autotransformer (see

189

Chapter 1). Such a device will enable the amplifier to deliver larger amounts of power into lower load impedances.

The input sensitivity of most amplifiers intended for professional use is on the order of 0.775 volts RMS for full output. If the amplifier has a potentiometer at its input, then the first stage cannot be overloaded. However, if the input has a buffer amplifier ahead of the potentiometer, the maximum input should be carefully noted and not exceeded.

A typical power output rating for an amplifier might read:

Output Power:	8 ohms:	4 ohms:
	225 W	380 W

What this states is that the amplifier can comfortably swing 42.4 volts RMS across a load of 8 ohms. The current corresponding to this is 5.3 amperes RMS.

If the amplifier were capable of delivering the same potential swing across 4 ohms, the resulting power would be 450 watts, and the current draw would be 10.6 amperes RMS. But for reasons of internal design and heat dissipation, the amplifier cannot do this. The device has been rated only at 380 watts into a 4-ohm load, and the potential swing must be limited to 39 volts RMS. The current which corresponds to this is 9.7 amperes RMS.

Although loudspeaker loads vary in impedance with frequency, it is convenient to label them in integral multiples or sub-multiples of 8 ohms, which result normally from running loads in series and/or parallel. Power output ratings related to different load impedances must be carefully noted, and in no case should an amplifier be allowed to produce a greater potential swing across a load than that corresponding to the rated power.

Bridging of Power Amplifiers:

In many cases, a stereo pair of amplifiers can be bridged, or operated with their outputs in series, to provide a better power match to a given load. For example, a pair of amplifiers rated at 200 watts into 8 ohms can be bridged to deliver 400 watts into 16 ohms. This is accomplished as shown in Figure 9-13B. An inverted drive is required at one of the amplifiers, and their output ground terminals must be common.

Manufacturers' specifications and instructions for bridging should be carefully noted and followed. Many an amplifier has been ruined through careless application here. Just as in the case with varying loads, manufacturers will specify maximum power ratings for various loads when the amplifier is operating in the bridged mode. Today, most amplifiers intended for bridged operation contain the required inverted drive internally, and an external switch activates the bridged operating mode.

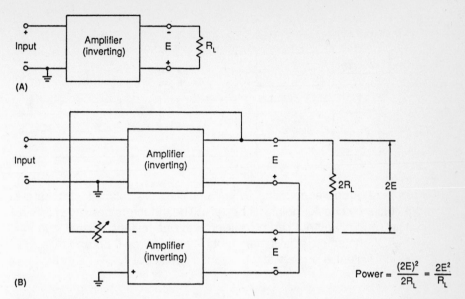

Figure 9-13. Bridging power amplifiers. **(A)** A single amplifier Power $= \dfrac{E^2}{R_L}$

(B) Two inverting power amplifiers bridged for doubled output power into *twice* the load impedance of a single amplifier; note that *twice* the potential across the load resistance produces *twice* the power as a single amplifier.

Amplifier Metering:

   It is superfluous to put output meters on power amplifiers that are intended for sound reinforcement use, as the operator of the system rarely will see them. It is, however, advisable for the amplifier to have some kind of indicator of signal output clipping, and for this function a simple LED indicator will suffice.

Amplifier Coupling:

   For certain installations requiring a very high degree of reliability, amplifiers may be coupled in such a way that they effectively act in parallel. Amplifiers cannot simply be operated in parallel. With even the slightest unbalance between them, one amplifier will attempt to drive the other as a load—and with disastrous consequences! A passive device known as an *amplifier coupler* places a reactive load between the two amplifiers to inhibit the mutual loading of the amplifiers. When the amplifiers are perfectly balanced, the reactive load is effectively out of the circuit. For any degree of imbalance, the reactive load serves to isolate the two amplifiers from each other. In the case of amplifier failure, the malfunctioning amplifier is isolated from the functioning

191

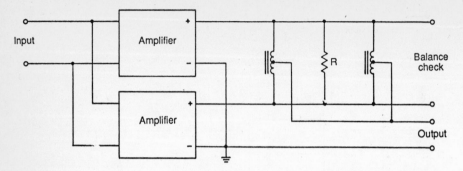

Figure 9-14. Amplifier coupling. When amplifiers are balanced, the output at the "balance check" point will be zero. The inductors are essentially at zero loss, and full output can be delivered at one-half the normal load impedance. If one amplifier fails, the output level will drop 6 dB and there will be a signal potential present at the balance check point.

one, and the level loss at the load will be no more than 6 dB. Details are shown in Figure 9-14.

WIRING CATEGORIES AND CONNECTORS

In general, in sound reinforcement design, we can consider the following wiring categories:
1. Signal:
   a. Microphone level (–30 dBm and lower)
   b. Line level (–30 dBm to +30 dBm)
   c. High level (+30 dBm and above, for loudspeaker power distribution)
2. AC powering
3. DC control signals

All of these categories should be well isolated from each other, but it is possible for DC control signal wiring to be run in proximity with line level signal wiring.

Wiring Classes:

Based upon national wiring code standards, signal wiring may or may not require protective conduit, depending on the powers involved. All sound contractors should be well versed in code details, and must be aware of any local municipal requirements that may affect wiring conventions.

Connectors:

Professional workmanship calls for reliable and secure connections between all elements in an audio chain. Specifically, RCA and most other "high fidelity" connectors are to be avoided, unless a piece of consumer gear is to be interfaced with a sound reinforcement system. Even then, only the best and most secure corrosion-proof connectors should be used.

The 3-pin XLR locking-type connector, in its many embodiments, is preferred for all microphone and portable line level interconnections. Figure 9-15A shows details of this connector.

Barrier strips are often used for permanent line level interconnections, and the leads should always be served with lugs so that short circuits do not inadvertantly occur. See Figure 9-15B for further details.

So-called phone plugs and jacks are common throughout the MI (Musical Instrument) field and have the advantage of being low cost. Their reliability is somewhat lower than the locking-type connectors.

Patch Bays:

A patch bay is a group of points where the inputs and outputs of elements in the audio chain can be accessed. While essential in a recording studio console, patch bays are a luxury in all but the most complex sound reinforcement systems. Patch bays afford flexibility in reconfiguring a system for a job different from its normal one, and the decision to implement it in this fashion should be based on experience.

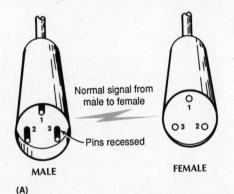

Normal signal from male to female

Pins recessed

MALE

FEMALE

(A)

Standard pin configuration:

Microphone	Line level
Pin 1 = Shield	Pin 1 = Shield
Pin 2 = Signal high	Pin 2 = Signal low
Pin 3 = Signal low	Pin 3 = Signal high

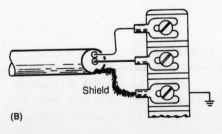

Shield

(B)

Figure 9-15. Preferred connectors. **(A)** The XLR-type connector. **(B)** A barrier strip with spade lugs on connecting wires.

Most sound reinforcement systems are designed to do just one job, and most of them are designed with inexperienced operators in mind. Thus, the specification of a patch bay in a sound reinforcement system should be weighed carefully.

More useful in a sound reinforcement system are test points, points where a signal may be monitored electrically without breaking circuit continuity. Equally useful is a switchable monitoring panel, which allows various points in a system to be checked for audio quality and electrical level. Figure 9-16 shows some of these options.

GROUNDING AND SHIELDING PRACTICE

System grounding is a complex subject, and there is no substitute for field experience in system construction and checkout. In this section, we can only touch the high points.

A distinction is made between signal ground and chassis ground; this leads to the notion of balanced and unbalanced lines.

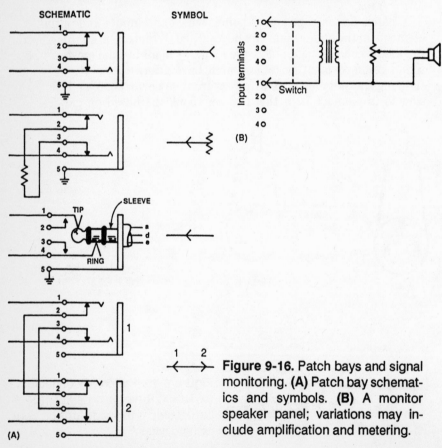

Figure 9-16. Patch bays and signal monitoring. **(A)** Patch bay schematics and symbols. **(B)** A monitor speaker panel; variations may include amplification and metering.

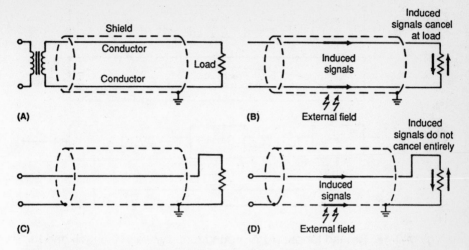

Figure 9-17. Balanced and unbalanced lines. **(A)** A floating balanced line. **(B)** Cancellations of induced signals in a balanced line. **(C)** An unbalanced or single-ended line; the shield is being used as one signal lead. **(D)** Incomplete cancellation of external fields in an unbalanced line.

Balanced and Unbalanced Lines:

In a balanced floating line, as shown in Figure 9-17A, two conductors are placed in a braided shield. The advantage here is that external electrical and magnetic fields induce equal signals in each of the conductors, and there is a net cancellation of the external field at the load. This is shown at B.

At C we show an unbalanced line, and at D we see how an external field is not completely cancelled at the load.

Obviously, balanced lines are essential, especially at microphone level, for maintaining quiet operation of a sound reinforcement system.

Ground Loops:

A ground loop exists when there is continuity in a ground path which surrounds a source of interference. Figure 9-18A shows an example of this. Electrical power mains can induce "hum" into the system, especially an unbalanced one, when this condition exists. The only remedy here is to break the continuity of the ground path, as shown at B.

Generally, the shield between active elements in an audio system is grounded only at one end, as shown in Figure 19. This procedure is usually sufficient to correct the problem. Furthermore, all of the shields are grounded at a single point, and that grounding point should be a very low resistance path to an earth ground. This is often referred to as "unipoint" grounding, and it should, in complex installations, extend to the AC power lines as well. In the most sophisticated systems, those

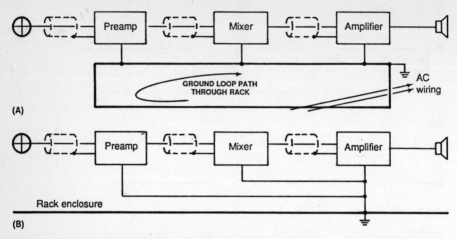

(A)

(B)

Figure 9-18. Ground loops. **(A)** A ground loop between the rack and the chassis. **(B)** Breaking the ground loop.

where only the lowest noise levels can be tolerated, the third wire, or chassis ground, in the AC plug will be individually wired directly back to a common ground point, bypassing the grounding path provided by the metal conduit. The term "technical ground" is also used to describe the low resistance earth ground to which all current drain paths are connected.

The reason for such elaborate systems is not only for the sake of low noise; safety is assured during equipment malfunction or inadvertant shorting to the chassis if these procedures have been followed.

One of the marks of a professional system designer and installer is that he understands the principles of grounding and shielding thoroughly.

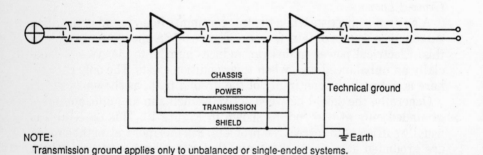

NOTE:
Transmission ground applies only to unbalanced or single-ended systems.

Figure 9-19. Unipoint grounding.

Line Conditioning:

A great deal of radio frequency (RF) interference finds its way into audio systems through the power mains. Heavy switching and certain kinds of lighting can produce spikes on the line. These, in turn, find their way through power supplies into the audio circuitry and manifest themselves as buzzing sounds. While proper shielding will take care of many of these problems, some degree of line conditioning may be necessary. Where possible, large audio installations should be powered from separate mains than those used for lighting and elevators, etc. If this is not possible, isolation transformers can filter out RF disturbances on the power line.

Good line potential stability or regulation is important as well, and there are various resonant transformer systems available to accomplish this. Experienced electrical contractors can be very helpful in these areas.

PRIORITY AND CONTROL SYSTEMS

Modern communications systems often call for considerable flexibility in routing signals. A typical paging and announcement system in a transportation terminal, for instance, may call for wide coverage for paging individual passengers, as well as small-area coverage for boarding information at individual gates. If we add to this the requirements of overall paging level control and recorded announcements, it is evident that we have the makings of a very complex system. There is little equipment specifically designed for signal switching and routing outside of that used in telephone interconnect practice. However, a number of experienced sound contractors and consultants have come up with their own proprietary designs.

Figure 9-20A shows details of an arrangement which allows certain inputs to take priority over other inputs. Paging type microphones normally contain manual switches to actuate them. When a microphone is not in use, it is shorted. When a microphone is actuated, the short is lifted and it is placed across the line; all microphones of lesser priority are shorted and cannot be used. Relays can also be used for the switching functions, and it is possible to use more than one input at each priority level. For further details, see reference (1).

At B, we show how the same approach can be applied to the requirements of zone paging and wide-area paging.

Another concern in signal distribution is to minimize the amount of wiring necessary to do a given job. The circuit shown in Figure 9-21 is useful in that it allows three wires to accomplish what would ordinarily require four wires. The circuit provides for background music with priority given to paging. Each source is individually powered for overall reliability.

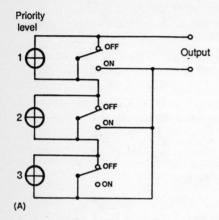

Figure 9-20. Priority systems. **(A)** Simple 3-level priority switching; activating and input disables all lower priorities. **(B)** Wide-area paging with priority over local zones.

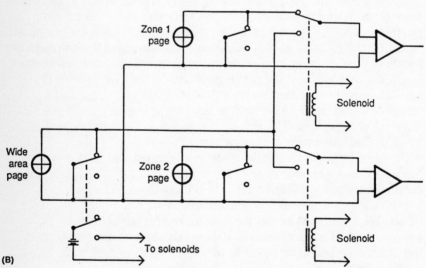

198

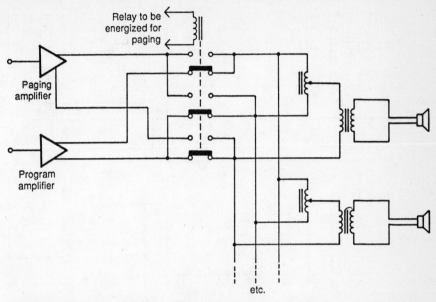

Figure 9-21. Priority switching at the loudspeakers (circuit courtesy of D. Klepper and P. Tappan).

Simplexing is a method of transmitting both signal (AC) and control (DC) along a single pair of wires. Transformers with center-tapped secondaries are required for these applications. Many functions can be handled in this way. For example, in Figure 9-22A, we show how the Power On Off status of a remote power amplifier can be indicated back at an earlier point where the signal to the amplifier originates.

The circuit shown at B allows a remote power amplifier to be turned on or off as required, and the circuit at C allows a remote signal to take priority over a local signal.

199

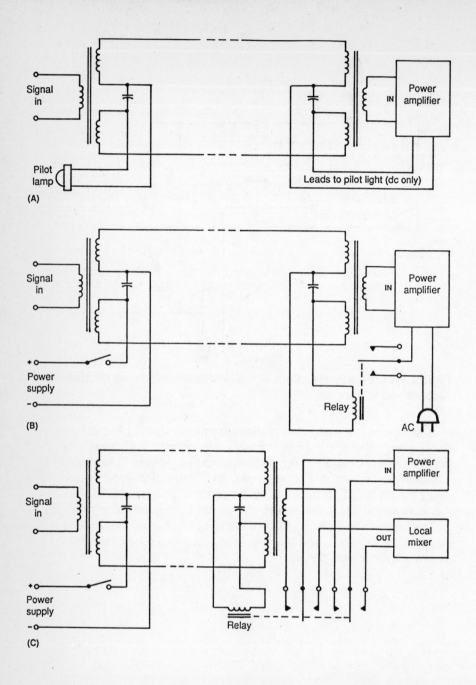

Figure 9-22. Examples of simplexing. **(A)** Remote indications of *on/off* status. **(B)** Remote turn on-off of a power amplifier. **(C)** A priority system.

Care must be taken that the current limits of the transformers used in simplexing are not exceeded. Generally, a limit of 100 milliamperes is assumed.

CONCLUSIONS

The lessons taught in this chapter are important ones; they should be mastered before proceeding forward in the book. We cannot stress enough the importance of good system construction and layout practice. In this chapter, we have merely touched upon this important subject, and the serious sound reinforcement system engineer would do well to apprenctice himself with experienced firms in the field so that he can learn proper practice directly from those who make it their livelihood.

CHAPTER 9:
Suggested Reading:
1. D. and C. Davis, *Sound System Engineering,* Howard W. Sams and Company, Indianapolis (1987).
2. J. Eargle and G. Augspurger, *Sound System Design Reference Manual,* JBL Incorporated, Northridge, CA (1986).
3. H. Tremaine, *Audio Cyclopedia,* Howard W. Sams and Company, Indianapolis (1969).

Concept of Acoustic Gain

INTRODUCTION

When we view the elements of a sound reinforcement system as a whole, we observe that there is an electro-acoustic feedback path around the system, as shown in Figure 10-1A. If there is a pulse of sound at the microphone, and if the electrical gain is high enough, the original pulse may be followed by one of equal amplitude, as shown at B. When this occurs, the system goes into sustained oscillation, or feedback. Two conditions are necessary for this to happen: the gain through the electro-acoustical loop must be *unity*, and the phase angle at the oscillating frequency must be *zero*. Anyone who has ever witnessed a system in feedback will not forget it, for it quickly brings into play the full power of the system.

It is the proper control of gain that makes a sound reinforcement system work in the first place, and in this chapter we will quantify the concepts of system gain for both outdoor and indoor cases and suggest rules for determining just how much gain a system should have.

AN OUTDOOR SOUND REINFORCEMENT SYSTEM

The basic elements of an outdoor sound reinforcement system are shown in Figure 10-2. We will make the following assumptions: that

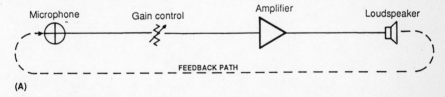

(A)

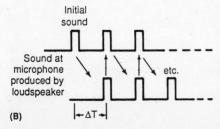

(B)

Figure 10-1. An outdoor sound reinforcement system: Electroacoustic relationships. **(A)** The feedback path through the acoustical environment. **(B)** Acoustical feedback, with a single pulse giving rise to a train of pulses.

there are no significant reflections (a free field is assumed to exist), and both microphone and loudspeaker are omnidirectional.

The acoustic gain of the system is defined as the increase in level a given listener will perceive when the system is turned *on*, as compared with the level he perceives when the system is *off*. Acoustic gain may thus vary with the distances between talker, microphone, loudspeaker, and listener. First, let us determine the various levels that exist when the system is off. Let the talker produce a level L at the microphone. Then, by inverse square law, the level at the listener will be:

$$\text{Level at listener} = L - 20 \log (D_o/D_s) \tag{10-1}$$

Now, let us turn on the system and increase the gain in the microphone-loudspeaker loop until we have just reached unity; that is, when the loudspeaker produces the same level L at the microphone that the talker produces. When this condition exists, we can calculate the level produced by the loudspeaker at the listener:

$$\text{Level at listener} = L - 20 \log (D_2/D_1) \tag{10-2}$$

Recalling that acoustical gain is defined as the difference between the levels given by Equations 10-1 and 10-2, we have:

$$\begin{aligned}
\text{Acoustical gain} &= L - 20 \log (D_2/D_1) - L + 20 (D_o/D_s) \\
&\quad 20 \log (D_1/D_2) + 20 \log (D_o/D_s) \\
&\quad 20 \log D_1 - 20 \log D_2 + 20 \log D_o - 20 \log D_s
\end{aligned} \tag{10-3}$$

Obviously, we cannot operate the system at the threshold of feedback, and it will be necessary to add a safety factor to the equation. Figure 10-3 shows the effect of operating a sound reinforcement system 3 dB

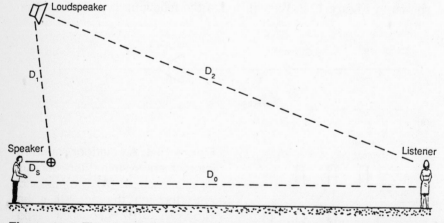

Figure 10-2. The physical relationships of an outdoor sound reinforcement system.

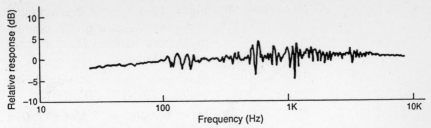

Figure 10-3. The electrical response of a sound system that is 3 dB below sustained acoustical feedback.

below the level of sustained oscillation. The response is quite irregular, and the peaks in the response represent potential feedback frequencies.

For most applications, a safety factor of 6 dB will suffice; therefore, we can write the gain equation in its final form:

$$\text{Maximum gain} = 20 \log D_1 - 20 \log D_2 + 20 \log D_o - 20 \log D_s - 6 \qquad (10\text{-}4)$$

We can make several observations at this point: that gain is independent of the level of the talker, that gain is proportional to the microphone-loudspeaker distance, and that gain is inversely proportional to the talker-microphone distance. These facts are of course intuitively obvious.

The Effect of Directional Microphones and Loudspeakers:

So far, we have considered only omnidirectional loudspeakers and microphones. Through proper application of directional loudspeakers and microphones, system gain potential can be increased. Let us examine the case shown in Figure 10-4A. Here, Equation 10-4 yields a maximum system gain of 7.5 dB.

Now, if we use a microphone whose directional pattern is as shown at B, and orient the microphone with respect to the loudspeaker as indicated, we will be able to pick up another 6 dB of gain potential giving 13.5 dB.

Similarly, we can take a loudspeaker whose directional pattern is as shown at C and orient it with respect to the microphones as indicated. In this case, we will be able to pick up another 6 dB gain potential.

As a practical matter, however, it is not possible to increase the gain potential of an outdoor system much beyond 6 dB through the use of directional components. The reasons for this have to do with the fact that, in the real world, ideal conditions never exist. Putting it another way, as we try to increase gain further, our assumptions must gradually become modified.

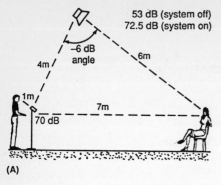

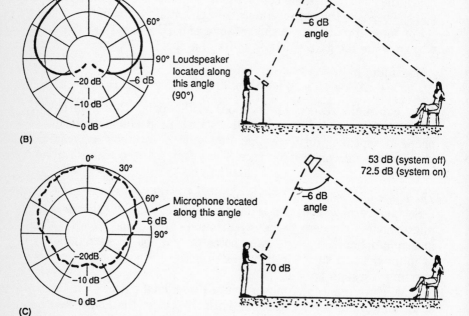

Figure 10-4. An outdoor sound system. **(A)** Assuming an ommni-directional loudspeaker, system gain potential is:

Maximum gain = 20 log(4) -20 log(6) + 20 log(7) -20 log (1) -6

Maximum gain = 12-15.5+17 -0 -6 = 7.5

(B) The polar plot of a microphone. **(C)** Orientation with respect to the microphone.

How Much Gain is Needed?

If a sound system has more than enough gain for normal speech reinforcement, then we can simply turn down the system to a comfortable point, and all will be well. In many cases, especially those where the ambient noise level may be high, a system may be marginal or deficient in its gain capabilities. What is needed is a method for determining beforehand just what the level requirements for a given system will be. One way of arriving at this is through the notion of *effective*, or *equivalent acoustic distance* (EAD). Referring to Figure 10-5, we can observe that a reinforcement system, in a manner of

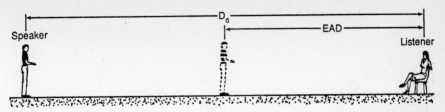

Figure 10-5. The concept of effective, or equivalent acoustic distance (EAD).

speaking, brings the listener and talker closer to each other by increasing the effective level of the talker as perceived by the listener. In quiet environments, adequate speech communication can be maintained over a distance of about 3 meters (10 ft), and we would not normally want a reinforcement system to produce levels any louder than those we would normally perceive at a distance of 3 meters from a talker. However, in a noisy environment, speech communication demands a much shorter distance, and we might wish to bring the talker to within a meter or so of the listener.

Determining the necessary gain of a sound reinforcement system requires that we calculate the inverse square relationship between D_O and EAD, as show below:

$$\text{Necessary gain} = 20 \log D_O - 20 \log \text{EAD} \qquad (10\text{-}5)$$

Referring back to our example, let us assume that an EAD of 3 meters will be sufficient. We see then there is a gain potential of 7.5 dB without the use of a directional microphone or loudspeaker. Calculating the necessary gain:

$$\begin{aligned}
\text{Necessary gain} &= 20 \log (7) - 20 \log (3) \\
&= 17 - 9.5 = 7.5 \, \text{dB}
\end{aligned}$$

Thus, our example barely meets its gain requirement!

Should we have specified a shorter EAD, then the system would not have had reserve gain to meet our requirement, and our recourse would be to change some parameter of the system. There are several. The first and most obvious change we could make would be to shorten D_S to 0.5 meter (20 in). This would allow us to pick up another 6 dB of gain potential. Alternatively, we could specify directional microphones and loudspeakers and pick up another 6 dB of gain reserve. In a later chapter on system intelligibility requirements, we will present data which will help the designer pick an appropriate EAD for various kinds of applications.

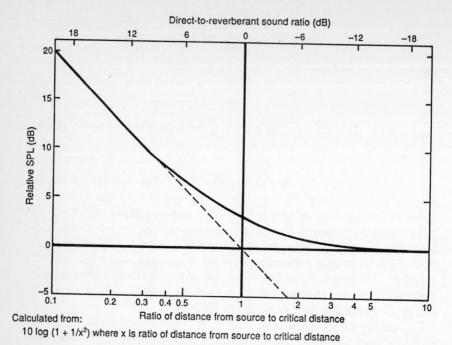

Figure 10-6. Relative SPL vs. distance from the source in relation to critical distance.

AN INDOOR SOUND SYSTEM

In analyzing the general case of an indoor reinforcement system, the attenuation with distance away from the talker or the loudspeaker must be calculated along its specific direct-reverberant attenuation curve. The general form of this curve is shown in Figure 10-6, and in working any examples, we must have an accurate knowledge of the directional characteristics of the microphone and loudspeaker, as well as the room constant R, all of these at the frequency at which we are making the system analysis. As a practical matter, gain analyses should be made at least at three frequencies. In addition to an analysis at 1 kHz, performance at 250 Hz and 4 kHz should be examined as well.

Let us work the example shown in Figure 10-7, restricting ourselves to a single frequency. First, we will calculate the critical distance for the talker. In the 1-to-2 kHz range, a talker will have a DI of about 3 dB, corresponding to a Q of 2; therefore:

$$D_C = .14\sqrt{(2)(111)}$$
$$D_C = 2 \text{ meters}$$

Now, we will calculate the critical distance for the loudspeaker.

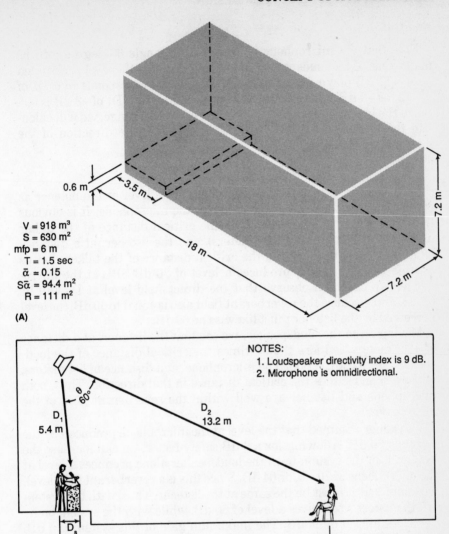

$V = 918\ m^3$
$S = 630\ m^2$
$mfp = 6\ m$
$T = 1.5\ sec$
$\bar{\alpha} = 0.15$
$S\bar{\alpha} = 94.4\ m^2$
$R = 111\ m^2$

(A)

NOTES:
1. Loudspeaker directivity index is 9 dB.
2. Microphone is omnidirectional.

(B)

Figure 10-7. An indoor sound reinforcement system. **(A)** Acoustical characteristics. **(B)** System relationships.

The radiation pattern of the loudspeaker is not uniform, and we must calculate the critical distance along each particular angle of interest. Along its major axis, the loudspeaker is assumed to have a DI of 9 dB, corresponding to a Q of 8. We may solve for critical distance as we did with the talker, or we can refer to Figure 2-25 and read it directly. From the graph, we read a value of 4 meters for critical distance along the major axis of the loudspeaker.

Note that the microphone is located at an angle 60 degrees off the major axis of the loudspeaker, and along this axis the level is assumed to be –12 dB down from the major axis. This corresponds to a DI of 9 – 12, or –3 dB. The value of Q corresponding to a DI of –3 dB is 0.5. Since the graph of Figure 2-25 does not cover this range, we will calculate the critical distance of the loudspeaker in the direction of the microphone as follows:

$$D_C = .14\sqrt{(.5)(110)} = 1 \text{ meter}$$

Next we will calculate the difference in the level at the listener as compared with the level of the talker at the microphone. It is obvious that the microphone is well within the critical distance of the talker (0.6 meters as compared to 2 meters), and the listener, at a distance of 12 meters, is well beyond the critical distance of the talker. Let us assume that the talker produces a level of 70 dB-SPL at the microphone. We can then observe that the direct field level at D_C will be 60 dB, and, because the reverberant field also is equal to 60 dB, the level perceived by the listener must likewise be 60 dB.

Moving on to the loudspeaker, we see that the listener, at a distance of 13.2 meters, is more than 3-times the critical distance of the loudspeaker in that direction. The microphone, at a distance of 5.4 meters, is more than 5-times the critical distance in that direction. Thus, both microphone and listener are well within the reverberant field of the loudspeaker.

We earlier assumed that the level at the microphone produced by the talker is 70 dB. Allowing for a 6-dB safety factor, we can increase the loop gain in the system until the loudspeaker alone produces a level at the microphone of 70 – 6, or 64 dB. Since this is a reverberant field level, we know that it must be the same at the listener. Thus, with the system off, the listener perceives a level of 60 dB, while with the system on he perceives a level of 64 dB. The maximum gain of this system is 4 dB, which is a small but perceptible amount.

Using a directional microphone with a pattern as shown in Figure 10-8, we can increase the gain potential of the system. The pattern we have shown has a DI of 5 dB on-axis, and this means that it rejects, by 5 dB, random signals as compared with on-axis signals. Using this microphone should allow us to increase the system's gain potential by about 5 dB.

However, we must consider the joint effect of the patterns of both loudspeaker and microphone. The loudspeaker is located in a direction about 70 degrees off the main axis of the microphone, and by inspection we determine that the DI in that direction is 3 dB. What this means is that the microphone still favors sounds arriving from that 70-degree direction some 3 dB more than it favors totally random sounds. Along

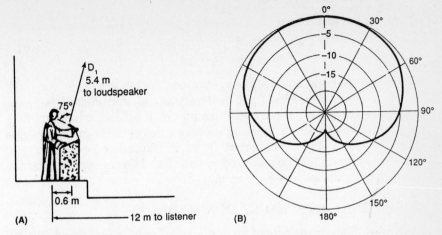

Figure 10-8. The effect of using a directional microphone in the system shown in Figure 10-7.

the same direction, the DI of the loudspeaker is –3 dB; therefore, the *joint DI* of the microphone and loudspeaker along their common axis is *zero*, corresponding to Q = 1.

We can now calculate the critical distance along that common path:

$$D_C = .14\sqrt{(1)\,(110)} = 1.5 \text{ meter}$$

Since the microphone is more than 3-times this distance of 1.5 meter, it is well within the reverberant field of the loudspeaker, and we should, in theory, be able to increase the system gain by 5 dB. However, more than 3 dB increase in gain in an indoor system is unlikely, due to the many discrete reflected paths between loudspeaker and microphone.

The Indoor Gain Equation:

As we have seen, calculating potential system gain for an indoor system can be a lengthy procedure; however, where both the listener and the microphone are in the reverberant field of the loudspeaker, and the microphone is in the direct field of the talker, then we can arrive at a simple gain equation. Assume that a talker produces a level L at the microphone. When the system is turned off, the level perceived by the listener will be:

$$L - 20 \log (D_{ct}/D_S),$$

where D_{ct} is the critical distance of the talker. Now, the system is turned on, and the gain is advanced until the loudspeaker produces a level at the microphone of $L - 6$ dB. At the same time, the level at the listener, due to the loudspeaker, will be $L - 6$ dB.

Taking the difference:

$$\text{Gain} = L - 6 - L + 20 \log (D_{ct}/D_s)$$
$$= 20 \log D_{ct} - 20 \log D_s - 6 \qquad (10\text{-}6)$$

Note that there is effectively only one variable in this equation, D_s, since D_{ct} is more or less fixed by the acoustical properties of the room.

It is important to realize that this equation is subject to the qualifications enumerated above. If the microphone lies in the transition region between direct and reverberant fields of the talker or microphone, then the system must be analyzed in detail, and levels calculated at the listener with the system both on and off.

MEASURING SYSTEM GAIN AND DELTA

Figure 10-9 shows details of how acoustical gain is measured. Boner[1] introduced the notion of system "delta" as a measurement of the maximum theoretical loop gain a system could produce. Both gain and delta can be measured using the same basic set-up as shown in the figure.

A small loudspeaker, usually a 125 mm (5 in) model, is located at the podium microphone at a distance similar to that of the talker. A pink noise signal, band-limited approximately to the octave centered at 1 kHz, is fed to the loudspeaker, and is adjusted to produce a level of about 80 dB-SPL at the microphone. This level must be carefully measured with a sound level meter, using its A-scale, with its microphone located just adjacent to the system's microphone. Then, with the system off, the sound level meter is located at a point well into the auditorium, and the level carefully noted.

The sound system is then turned on, and the gain is advanced until the system is just below the point of sustained oscillation. The sound

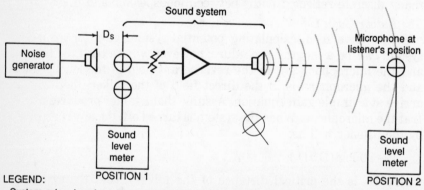

LEGEND:
System gain = Level, position 2, system ON − level, position 2, system OFF
SYSTEM Δ = Level, position 2, system ON − level, position 1, system OFF

Figure 10-9. Measurement of system gain and Delta.

level meter is then taken back into the house, and the level carefully noted.

The system gain of course is the difference in levels noted at some point out in the house, while the system delta is the difference between the level in the house with the system on and the level at the microphone with the system off.

Gain is a function of D_S, while delta is not. Theoretically, with precise control of many variables and through the use of narrow-band equalization, a delta of unity can be approached.

CHAPTER 10:
Reference:
1. C. and R. Boner, "The Gain of a Sound System," *J. Audio Eng. Soc.*, Vol. 17, No. 2 (1969).
Additional Reading:
1. D. and C. Davis, *Sound System Engineering*, Howard W. Sams and Company, Indianapolis (1987).
2. J. Eargle and G. Augspurger, *Sound System Design Reference Manual*, JBL Incorporated, Northridge, CA (1986).

System Equalization Practice

INTRODUCTION

Although sound system equalization has been practiced since the thirties, it was not until the sixties that Paul Boner established the basis for its application on a large scale. Altec's introduction of "Acousta-Voicing" in the mid-sixties further broadened the basis of equalization in general sound contracting work, and it is unusual today to find a professional sound reinforcement or playback system which has not benefitted from some degree of equalization.

Boner's unique contribution was the increase in system gain through the "notching out" of potentially troublesome feedback frequencies through the use precisely tuned narrowband filters. Playback system equalization, on the other hand, makes use essentially of broadband, gently sloped filters in order to smooth out the system's power response.

Thus, it has become convenient to speak of narrowband and broadband equalization, realizing that they are quite often used in conjunction with each other. While the process is generally referred to as equalization, the devices that are used are normally called filters, inasmuch as their specific action has been associated for years with the removal, or "filtering out," of unwanted portions of the frequency spectrum.

FILTER CHARACTERISTICS

The filter designs preferred for sound reinforcement are minimum phase; that is, the phase shift associated with a given amplitude correction is minimal, without exhibiting any form of all-pass group delay. While practice varies according to the preference of the system designer, those filters which provide only a cut or attenuation action normally used. The reason here is simply that such devices are generally easier to work with than those that provide both cut and boost.

Broadband Devices:

The most general form of filter intended for broadband equalization has attenuation functions spaced on ISO one-third octave center fre-

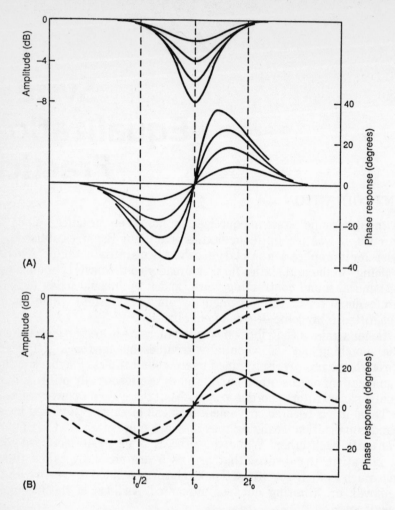

Figure 11-1. Combining type filter for broadband equalization; a single section is shown at *A*. Three adjacent sections are shown at *B*. (Note that the three act as one broader section.)

quencies covering the range from about 40 Hz to 16 kHz. Figure 11-1 shows typical amplitude and phase characteristics of these devices, and Figure 11-2 shows photographs of some commercial units.

Figure 11-3 shows characteristic response curves for broadband equalization of a system. Both amplitude and phase have been plotted. Note that the various filter sections interact to provide a smooth overall curve. For this reason, these filters are often referred to as "combining-type" filters.

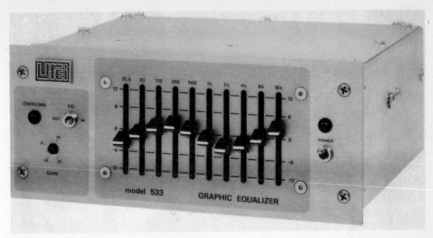

(A)

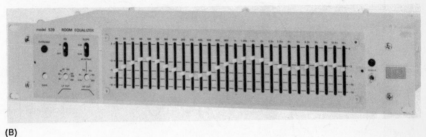

(B)

Figure 11-2. Commercial units for broadband equalization. (UREI Photo).

In general, the equalization range of most commercial devices is far in excess of that which most systems will require. There is, in fact, a movement toward less broadbanding of system response as more attention is paid to smooth power response in specifying LF and HF system components. If only a gentle amount of broadbanding is required, then one-octave wide filters will do the job nicely.

Narrowband Devices:

Narrowband devices must be individually tuned to the feedback frequencies they are intended to control. Figure 11-4 shows how passive L-C sections are inserted between amplifiers for notching action at particular frequencies. Figure 11-5 shows the phase and amplitude characteristics for a typical narrowband section.

Figure 11-6 shows the amplitude response of four narrowband sections in series, and Figure 11-7 shows a photograph of a typical narrowband filter.

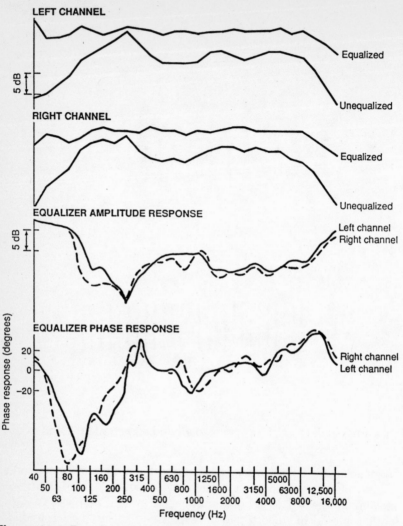

Figure 11-3. Broadband equalization of a 2-channel monitoring system.

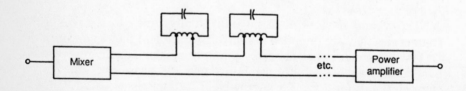

Figure 11-4. Implementation of narrowband filters.

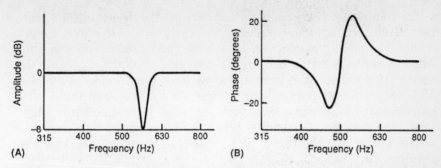

(A)

(B)

Figure 11-5. Amplitude and phase response for a typical narrowband filter set for 8 dB of attenuation.

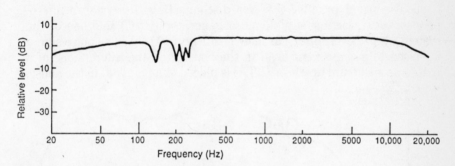

Figure 11-6. Four narrowband section in series.

Figure 11-7. Commercial narrowband filters. (UREI Photo).

219

The combination of broadband and narrowband devices results in an overall electrical equalization curve such as that shown in Figure 11-8A, while the overall system acoustical response associated with this filtering action is shown at B. In this case, the system was a large-scale speech reinforcement system installed in a large hotel ball room. Note that the four narrowband notches do not influence the broadband nature of the overall acoustical response; their function is solely to reduce feedback tendencies. Since the narrowband dips are well within the critical bandwidth, as discussed in Chapter 3, the dips will not be audible as such.

One-third octave filters can be used to a limited extent in reducing feedback tendencies after a system has been broadbanded. The problem, however, is that fixed one-third octave centers can only approximate the actual feedback frequencies.

TYPICAL PROCEDURES

In equalizing a reinforcement system, broadbanding is always done first. The usual practice involves disconnecting the system's microphones and inserting a pink noise generator (PNG) into one of the system's high-level inputs, as shown in Figure 11-9A. The noise signal is raised to a convenient level in the house, and the microphone of a real-time spectrum analyzer (RTA) is placed at a point out in the house,

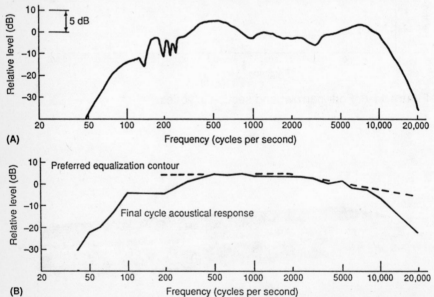

Figure 11-8. The combination of broadband and narrowband filtering. **(A)** Electrical response of both filter types. **(B)** Resulting acoustical response.

220

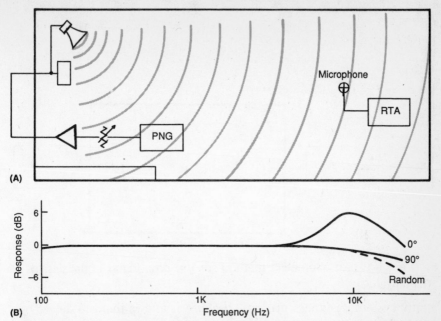

Figure 11-9. Broadband equalization of a sound reinforcement system. **(A)** Set-up in the house. **(B)** Ideal measuring microphone characteristics.

usually about half or two-thirds of the way back. In this position in the room, the microphone is well into the reverberant field of the loud-speaker. Thus, the microphone will pick up the integrated output of the loudspeaker, and the measured level will, with respect to frequency, correspond to the power response of the loudspeaker system.

Microphone characteristics are very important. It is commonly agreed that a flat random incidence microphone should be used for equalization work, and this calls for a microphone with response as shown in Figure 11-9B. In measurements made in the reverberant field, the microphone's orientation is not critical. If, as in the case of equalizing a monitoring system, the measurements are made in a location where the direct field may be significant, then the micro-phone should be oriented at 90 degrees to the source, since the 90-degree response of the microphone closely approximates the random incidence response.

The RTA registers the acoustical level in each of its displayed bands. A one-third octave RTA is required if one-third octave filters are used.

An alternative method is shown in Figure 11-10. Here, a one-third octave wave analyzer feeds only one band at a time into the system, and a standard SLM can be used for making measurements. While this method takes somewhat longer than that shown in Figure 11-9A,

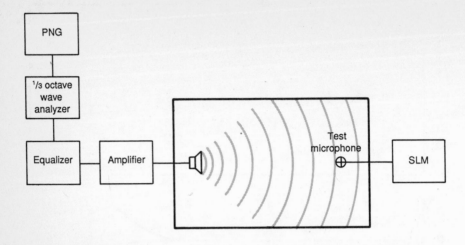

Figure 11-10. An alternative method for the broadband equalizing of a sound system.

many system designers prefer it because it allows them to detect by ear certain acoustical anomalies that may be present.

Preferred House Curves:

The term house curve refers to the frequency contour that the system's response is shaped to match. Early in his work in reinforcement system equalization, Boner established the curve shown in Figure 11-11A as an ideal one for preserving a natural quality in amplified speech.

Why is the curve not flat? The HF rolloff has to do largely with the directional characteristics of the HF devices which were prevalent during the sixties. At Figure 11-11B we show the DI of a typical HF radial or multicellular horn. Since we are effectively equalizing the system's power response in the house, we note that the sum of the power response (house curve) and the DI will effectively give us the on-axis direct field response of the system. As can be seen, the preferred house curve and the DI combine to give us a nearly flat direct field. Since the direct field reaches the listener's ears first, it has, through the precedence effect, a profound influence on the quality of the perceived sound. Boner's judgement was a subjective one, and it was fundamentally correct as regards the integrity of the direct sound field.

In recent years, the use of constant coverage HF horns has made it possible to modify the preferred house curve by maintaining it flat further out in frequency.

Whether one chooses to keep the preferred Boner house curve or not, the use of constant coverage devices certainly results in more consistent reverberant and direct field levels, as shown in Figure 11-12.

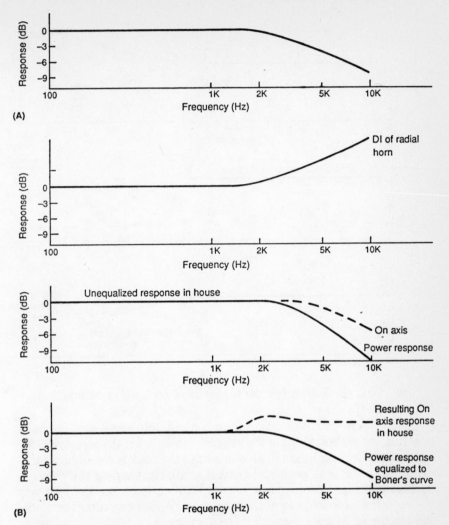

Figure 11-11. Preferred house curves. **(A)** Boner's preferred curve. **(B)** Boner's preferred curve produces nearly flat on-axis response in the house.

The Equalization Process:

The process of broadband equalization has been made to look easier than it really is, and many people race through the process in too short a time. Here are some general recommendations:

1. Before proceeding with equalization, check the coverage in various parts of the house with the RTA. Any differences noted here will be the same after equalization as before, and of course any major coverage problems noted here should be corrected before equalization begins.

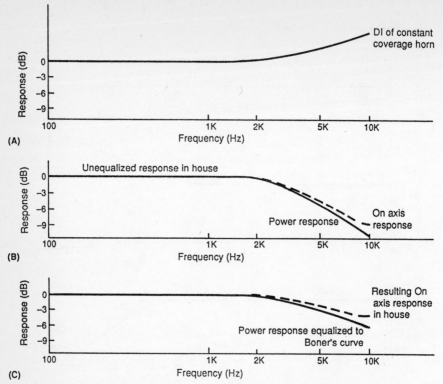

Figure 11-12. On-axis and power response of constant-coverage horns.

2. Use no more equalization than is absolutely necessary. Begin by spotting the highest point in the response; then, lower the corresponding control on the filter by small amounts until the peak is removed. At that point, take the next peak and treat it similarly, shaping the response to match the preferred house curve. Without question, the best way to learn the process is to observe an experienced consultant or sound contractor as he goes through these operations.

3. If the system is to be used primarily for speech reinforcement, then the frequency response extremes can be rolled off (below 100 Hz and above 8 kHz).

4. It is best for the beginner to go through the process several times, noting all filter settings after each trial. When final settings have been arrived at, it will be noted that system gain was reduced somewhat through the mid-band as a result of the various attenuation settings. At this point, the input and output acoustical levels, as well as the metering level, should be reset as required.

5. As tricky as broadband equalization is, narrowband equalization is even more so. It is strongly recommended that it be learned from an

experienced practitioner. The general procedure is carried out after broadbanding has been completed. With the system microphones in place, the relative gains should be established, with the system's overall gain set at a point well below feedback. Then, actuating the microphones individually, system gain is raised until feedback is just noticeable. This will take the form of non-sustained "ringing," most often in the frequency range of 100 to 1200 Hz.

A narrowband filter section is then adjusted so that its resonance frequency is the same as the ringing frequency, and no more attenuation should be inserted than is necessary to stop the ringing. It is ideal if the filters can be located in the specific microphone channel where they are used, as shown in Figure 11-13. It is not practical to introduce narrowband equalization into the input channel for a hand-held, roving microphone.

The narrowband process may be carried up to about four filter sections per microphone. Beyond that number, the increase in gain potential will diminish rapidly. In most cases, up to 3-5 dB acoustical gain increase may result from the use of narrowband filters.

6. The elevated gain setting made possible through the use of narrowbanding should be carefully monitored for overall stability under a variety of operating conditions. Environmental changes, such as temperature, relative humidity, and audience size will modify to some degree the precise nature of the electroacoustical feedback path. If the number of filters per microphone channel has been kept fairly low, and if the increase in gain due to filtering is no more than, say, 3-5 dB, then the effect of environmental changes will be minimal. The microphones may be changed, but pattern type and sensitivity should be kept the same.

Note should be carefully made of the maximum gain settings that

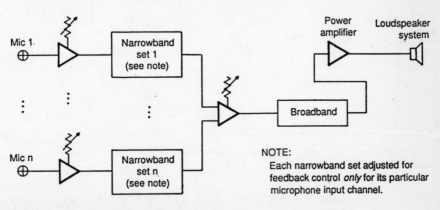

Figure 11-13. The ideal implementation of narrowband filters.

the equalized channels can be operated at before feedback becomes evident.

LIMITS ON THE EQUALIZATION PROCESS

It is important to know what the equalization process can and cannot accomplish, and the following comments sum up our observations:

1. Equalization cannot correct for irregular coverage of a loudspeaker system or for the effects of LF standing waves in a room. Such variations will be as evident after equalization as they were before, and, if anything, the performance of the system will be worse after equalization.

2. Equalization can correct for moderate variations in the reverberant field level at mid- and high-frequencies due to boundary absorption in the space.

3. Equalization can correct for minor power response deficiencies in systems that have been otherwise designed for constant coverage over the entire frequency range. However, since many power response problems can be related to insufficient specification of hardware and power allotment, it is strongly recommended that problems in power response be corrected in the specification of loudspeaker components themselves and their respective power amplifiers.

4. Narrowband equalization should be used solely for the control of feedback modes in a complete electroacoustical system and never for the control of response peaks in transducers.

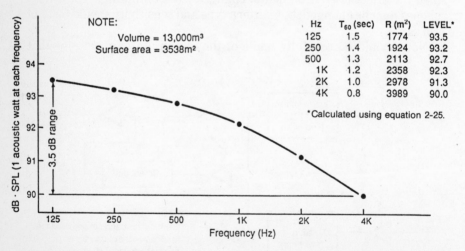

Figure 11-14. Reverberant levels in a typical auditorium for acoustic power input of one watt at octave centers from 125 Hz to 4 kHz.

FLAT POWER RESPONSE IN THE
TYPICAL ACOUSTICAL ENVIRONMENT

It is relatively easy to design loudspeaker systems that exhibit quite flat power response over a very wide frequency range. When such a system is placed in a well-designed acoustical environment, the need for broadband equalization will be minimal. Figure 11-14 shows the variation in reverberant level in a typical large auditorium for power input of one acoustic watt at each octave center from 125 Hz to 4 kHz.

In general practice, some equalization would be useful in the LF boundary dependent region, where loudspeaker loading and boundary reflections would result in some degree of variation in LF output. Over the bulk of the MF range, the acoustical level would be determined very largely by the room constant, while in the range above 4 kHz the level would fall off more rapidly due to air absorption. Note that in this example the amount of equalization required to match the Boner curve is minimal.

CHAPTER 11:
Suggested Reading:

Books:
1. D. and C. Davis, *Sound System Engineering*, Howard W. Sams and Company, Indianapolis (1987).
2. J. Eargle, *Handbook of Recording Engineering*, Van Nostrand Reinhold, New York (1986).

Articles:
1. C. P. and C. R. Boner, "Minimizing Feedback in Sound Systems and Room Ring Modes with Passive Networks, *J. Acoustical Soc. Am.*, Vol. 37, p. 131 (1965).
2. C.P. and C. R. Boner, "A Procedure for Controlling Roon-ring Modes and Feedback Modes with Narrow-band Filters," *J. Audio Eng. Soc.*, Vol. 3, No. 4 (1965).
3. J. Eargle, "Equalizing the Monitoring Environment," *J. Audio Eng. Soc.*, Vol. 21, No. 2 (1973).
4. J. Eargle and M. Engebretson, "A Survey of Recording Studio Monitoring Problems," *Recording Engineer/Producer*, Vol. 4, No. 3 (1973).
5. R. Schulein, "In Situ Measurement and Equalization of Sound Reproduction Systems," *J. Audio Eng. Soc.*, Vol. 23, p. 178 (1975).

System Intelligibility Criteria

INTRODUCTION

The intelligibility of speech is of paramount importance in public meeting places and auditoriums, and sound reinforcement systems should always be designed with the goal in mind of increasing intelligibility over that afforded by unamplified speech. The most important factors in determining speech intelligibility are:

1. *Speech level and signal-to-noise ratio.* Speech can be understood over a wide range of levels; however, at lower levels, the signal-to-noise ratio must be of the order of 25-30 dB if speech is to be clearly understood. At higher speech levels, a lesser signal-to-noise ratio will often suffice.

2. *Reverberation time.* If the reverberation time in the 500 Hz to 2 kHz range is of the order of 1.5 seconds or less, then it will not decrease intelligibility. In fact, a reverberation time of 1.5 seconds or less is generally beneficial, since it increases the level of speech without interfering with the articulation of individual speech syllables. (The nature of the reverberation may be critical; in particular, strong discrete echoes will make for poor intelligibility if they are noticeably displaced from the direct speech sound source.)

3. *Direct-to-reverberant ratio.* For reverberation times in excess of 1.5 seconds, the overhang of sound tends to blur speech. In a sense, the reverberation behaves like a kind of noise signal, one that rises and falls with the level of speech itself.

4. *Subjective considerations.* There are both good and bad listeners and talkers. Given the same acoustical environment, an experienced talker will adjust his delivery to the conditions at hand, and he will be more clearly understood than an inexperienced talker. Likewise, an attentive listener with normal hearing will be at an advantage over an elderly person with some degree of hearing loss.

MEASUREMENT OF SPEECH INTELLIGIBILITY

The traditional measure here is syllabic articulation testing. In such tests, a reader calls out a list of unrelated monosyllabic words, and listeners at various parts in an auditorium write down the words as they perceive them. The articulation score is simply the percentage of syllables correctly identified. If a given listener correctly identifies 85%

of the total number of syllables, then he will be able to understand normal speech in the testing environment with an accuracy of 97% or higher, due to the contextual nature of speech. If his articulation score is 75%, then he will be able to understand approximately 94% of the words in normal speech. This latter condition may be generally satisfactory, but some extra measure of attentiveness on the part of the listener may be necessary as well.

ESTIMATING SPEECH INTELLIGIBILITY

What the designer of sound reinforcement systems requires is a method of estimating speech intelligibility before a system is designed. Armed with such knowledge, he can determine just what kind of system may be best for a given environment.

Articulation Index (AI)

One of the earliest methods of determining the intelligibility of a transmission system is through the calculation of AI, as developed by French and Steinberg[1] and later modified by Kryter[2]. In calculating AI, the spectrum of ambient noise in the listening space must be measured, either on octave bands or one-third octave bands, over the range from 250 Hz to 4 kHz. Speech peaks are also measured over the same bands, and the two are compared at each frequency center. Figure 12-1 shows a method for arriving at the contribution of each octave band in determining AI, as suggested by Smith[3]. In using this graph, octave band RMS speech-to-noise ratios are measured and are individually weighted for their contribution to the AI over the normal 30-dB dynamic range of speech. Since it is easier to measure the RMS value of speech signals

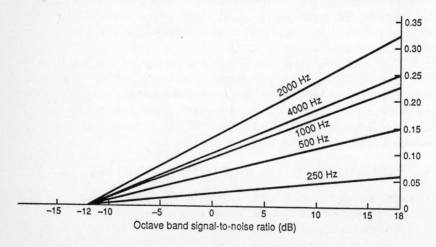

Figure 12-1. Calculations of the Articulation Index (AI).

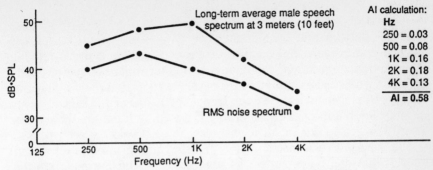

Figure 12-2. Sample speech and noise spectra; AI calculation.

than peak values, it is this signal-to-noise ratio that is plotted in the graph. The assumption is made that peak levels of speech are some 12 dB higher than the RMS levels.

As an example of how to calculate AI, let us assume noise and RMS speech spectra as shown in Figure 12-2. Each band is individually observed, and its RMS signal-to-noise ratio entered into the graph of Figure 12-1. A line is drawn upward from this value until it intersects the diagonal line representing the octave band of interest. Then, a line is drawn to the right, intersecting the vertical axis. Each contribution, as measured on this axis, is noted, and they are summed to give the AI.

AI calculations are most applicable to non-reverberant conditions, such as those found in open-plan offices and many paging applications. There is generally excellent agreement between AI calculations and articulation tests made under these acoustical conditions.

Figure 12-3 compares AI with several measures of speech intelligibility. In general, an AI of 0.3 or greater indicates that general speech intelligibility will be adequate.

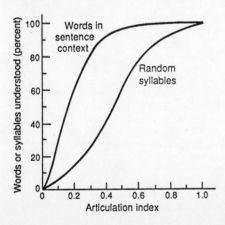

Figure 12-3. A comparison between AI and syllabic testing.

Peutz's Articulation Loss of Consonants

AI calculations have been found to give erroneous results if they are made under reverberant conditions. Peutz[4] suggests a measurement of the articulation loss of consonants as a determinant of speech intelligibility which takes into account reverberation time, noise, and the direct-to-reverberant ratio in the frequency range of 1-2 kHz. We can simplify Peutz's method considerably if we apply it only in those cases where the noise floor is sufficiently low (30 dB below speech peaks). In this case, noise is not a factor in determining intelligibility, and Peutz's data can be replotted, as suggested by Augspurger[5], into the form shown in Figure 12-4.

In using this data, the reverberation time in the 1-2 kHz octave is measured or calculated, and the direct-to-reverberant ratios at various parts in an auditorium are measured or calculated. The data is entered into the graph, and an estimate can be quickly made of overall system intelligibility. Peutz states that this method has a limit in accuracy of about 10%, and it is for this reason that the data of Figure 12-4 is broken down into only four broad zones of intelligibility.

The preceding methods are easy to work with, and they can be implemented while a sound reinforcement system is still on the drawing board if reasonable noise and reverberation time estimates can be made. There are two other methods of estimating speech intelligibility, but they require actual measurements on site. We will discuss them briefly.

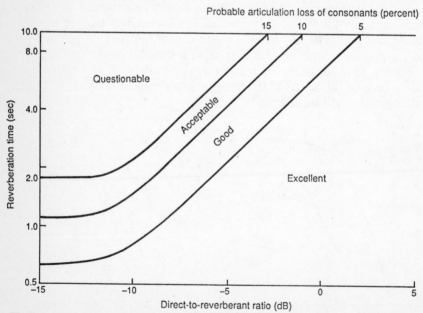

Figure 12-4. Articulation loss of consonants.

Lochner and Burger's Signal-to-Noise Method

Lochner and Burger[6] have determined that sound arriving within a certain interval after the receipt of direct sound is integrated by the ear and is useful. All sounds arriving after that time are considered as noise. The integration time is 95 msec, and the expression for useful sound energy is:

$$\text{Useful energy} = 10 \log \int_{t=0}^{95\,\text{ms}} \alpha\,(p,t)\,p^2(t)\,dt$$

In this expression, $\underline{\alpha}$ is a fraction of delayed sound, integrated taking into account the direct sound and the delay time. $p(t)$ is the instantaneous sound pressure.

The signal-to-noise ratio is defined as:

$$S/N = 10 \log \frac{\displaystyle\int_0^{95\,\text{ms}} \alpha\,(p,t)\,p^2(t)\,dt}{\displaystyle\int_{95\,\text{ms}}^{\infty} p^2(t)\,dt}$$

Figure 12-5 shows the relationship between the S/N calculation and speech intelligibility. Clearly, the method is an accurate one, but its implementation is quite complex.

Modulation Transfer Function (MTF)

Houtgast and Steeneken[7] have proposed a method for measuring the effects which reverberation and noise have on signal integrity. Their method makes use of a test signal which is amplitude modulated and then reproduced in a room. The effects of reverberation and noise fill

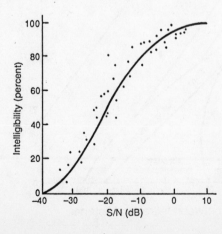

Figure 12-5. Signal-to-noise (S/N) vs. speech intelligibility; the curve gives the best fit to data points (after Smith).

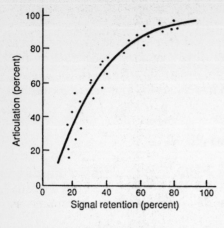

Figure 12-6. MTF vs. speech intelligibility; the curve gives the best fit to data points (after Smith).

in portions of the modulation envelope, and the recovered signal can be compared with the original. In practice, a number of modulation rates are used at a number of one-third octave bands. Figure 12-6 gives an indication of the accuracy of the method, stated by Houtgast and Steeneken to be within 10%.

In recent years, the RASTI (Rapid Speech Transmission Index) measurement method has evolved out of the work of Houtgast and Steeneken.

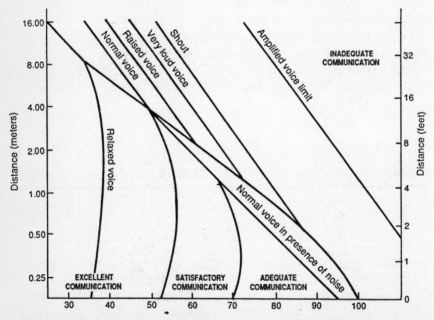

Figure 12-7. Permissible distances between talker and listener.

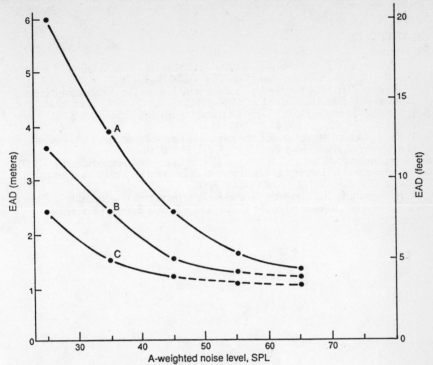

Figure 12-8. EAD vs. A-weighted noise level.

PERMISSIBLE TALKER-LISTENER DISTANCES FOR SPEECH COMMUNICATION

The graph shown in Figure 12-7 shows permissible talker-listener distances as a function of noise level. The assumption is made that the listener and talker are not facing each other. If they are facing each other, then the noise level may be increased by 5 dB.

This graph clearly indicates the tendency of talkers to raise their speech level in the presence of noise, and continued communication under elevated noise conditions is a strain on both talker and listener.

As an aid in determining operating levels for sound reinforcement systems, the data of Figure 12-8 will be useful. Here, we have plotted workable speech levels, as a function of noise level, for a raised voice (Curve A), a normal voice (Curve B), and a lowered voice (Curve C). In plotting these curves, the peak speech levels at a distance of one meter for the three types of talkers were set at 70 dB, 65 dB, and 60 dB, respectively. At lower noise levels, a 25 to 30 dB signal-to-noise ratio is maintained, while at higher noise levels the signal-to-noise ratio is allowed to decrease.

This data will be useful in establishing a value for EAD, as discussed in Chapter 10.

235

CHAPTER 12:

References:

1. N. French and J. Steinberg, "Factors Governing the Intelligibility of Speech Sounds," *J. Acoustical Soc. Am.*, Vol. 19, pp. 90-119 (1947).
2. K. Kryter, "Methods for the Calculation and Use of the Articulation Index," *J. Acoustical Soc. Am.*, Vol. 34, p. 1689 (1962).
3. H. Smith, "Acoustic Design Considerations for Speech Intelligibility," *J. Audio Eng. Soc.*, Vol. 29, pp. 408-415 (1981).
4. V. Peutz, "Articulation Loss of Consonants as a Criterion for Speech Transmission in a Room," *J. Audio Eng. Soc.*, Vol. 19, No. 11 (1971).
5. G. Augspurger, *JBL Sound Workshop*, JBL Incorporated, Northridge, CA (1977).
6. J. Lochner and J. Burger, "The Influence of Reflections on Auditorium Acoustics," *Sound Vibration*, Vol. 4, pp. 426-454 (1964).
7. T. Houtgast and H. Steeneken, "Envelope Spectrum and Intelligibility of Speech in Enclosures," (presented at the IEEE-AFCRL 1972 Speech Conference).

Central Loudspeaker Arrays for Speech Reinforcement

INTRODUCTION

Under appropriate acoustical conditions, a properly designed central array will result in very natural reinforcement of speech. It is generally the system of choice for auditoriums and houses of worship, where the spoken word is all-important. The preferred location for the central array is above the center at the front of the space aimed downward toward the listeners. For example, an excellent location in a theater would be just in front of the proscenium center.

For best results, the length of the room should not exceed about three-times the height of the array above the floor. If the ceiling is low relative to the length of the room, then those listeners toward the front of the room may perceive the sound from the array as too loud. In these cases, some kind of distributed system will be best. Usually, a hybrid system, one making use of a central array to cover the front portion of the space, with additional delayed loudspeakers for the rear of the room, will be the choice in a low-ceiling auditorium.

COVERAGE REQUIREMENTS

As a general rule, central arrays are custom designed for the space they are to operate in. The interrelation between coverage requirements and intelligibility requirements is important, and the flow diagram presented in Figure 13-1 outlines the design steps conceptually. What the diagram states in effect is that, as each loudspeaker element is added to the array, an accounting of direct-to-reverberant ratios in the seating area should be made and the effect on system intelligibility assessed.

Foreshortened Coverage Angles

When seen in plan view, the nominal coverage angle of a HF horn, as it appears in the drawing, will be a function of the elevation angle of the horn. Let us take the case of a horn whose nominal horizontal coverage angle is forty degrees at the –6 dB points. When the horn is aimed parallel to the floor (elevation angle zero), then the apparent

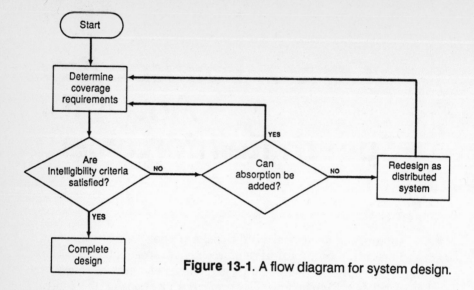

Figure 13-1. A flow diagram for system design.

angle in the drawing will be forty degrees, as shown in Figure 13-2. As the horn is progressively aimed downward, the apparent coverage angle as seen in plan view becomes wider, as shown at B. Of course, the actual radiation angle of the horn remains forty degrees, but when seen foreshortened, the angle appears to splay outward.

The equation which gives the foreshortened value of the angle is:

$$\theta' = 2 \arctan \left(\frac{\tan (\theta/2)}{\cos (\phi)} \right) \qquad \text{13-1}$$

In this equation, θ is the nominal coverage angle of the horn, ϕ is the elevation angle, and θ' is the apparent coverage angle as seen in plan view.

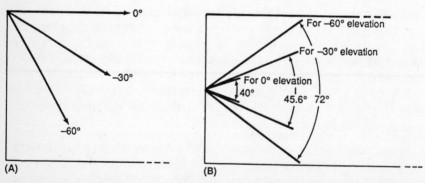

Figure 13-2. Foreshortened views of coverage angles. **(A)** Side elevation view. **(B)** Top view.

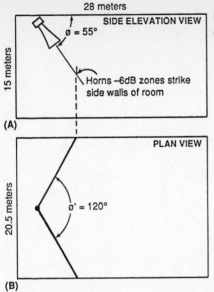

Figure 13-3. Foreshortened horn coverage angles, 90° X 40° horn.

With this equation, a designer can determine precisely the coverage on the floor of a horn tilted downward.

In the example shown in Figure 13-3, a 90-by-40 degree horn is tilted downward 55 degrees. It is clear from the construction lines in the views at A and B that the horizontal –6 dB points of the horn's coverage are well up on the side walls of the room. For this application, the horn is too wide in the horizontal plane, and it should be replaced with a 60-by-40 degree horn.

Horn directional data in isobar form is very useful in laying out systems in this manner. Figure 13-4 shows an example of isobars for a 90-by-40 horn. Note that the general shape of the isobars is elliptical and not rectangular, as is often assumed.

A Design Example

Figure 13-5A and B show side elevation and plan views of a typical house of worship. The length of the space demands some kind of long-throw coverage for the rear and balcony areas. Such a device will not provide adequate coverage for the front of the room, so at least one additional device will be needed.

Examining the side elevation of the space, a nominal 20-by-40 degree horn will be aimed toward the front of the balcony at an elevation angle of –25 degrees. We plot this directly as shown at A. Note that the 20-degree vertical coverage angle will provide good coverage of the balcony and a portion of the main floor below it. Using equation 13-1, we calculate the apparent horizontal coverage angle as seen in plan

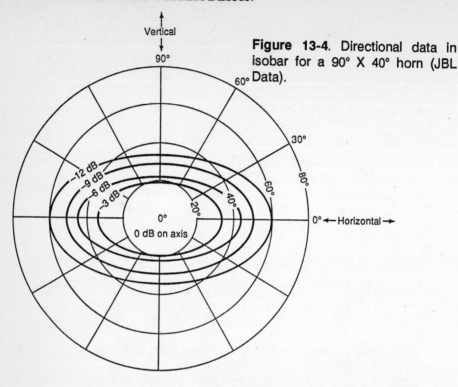

Figure 13-4. Directional data in isobar for a 90° X 40° horn (JBL Data).

view, and we draw this in the view at B. The value θ' of 44 degrees is shown.

Thus far, we have good coverage of the back of the room, and it will be instructive to see what the direct-to-reverberant ratio at the rear of the room will be with only this one loudspeaker element in place. To do this, we take into account the DI and sensitivity of the HF horn:

$$DI = 16.5 \text{ dB}$$
$$\text{Sensitivity (1 W at 1 m)} = 118 \text{ dB-SPL}$$

Using equation 4-5, we calculate the efficiency of the horn-driver combination:

$$10 \log \text{Eff} = 118 - 109 - 16.5$$
$$= -7.5 \text{ dB}$$

The efficiency is then taken as:

$$\text{Efficiency} = 10^{\frac{-7.5}{10}} = 0.18, \text{ or } 18\%$$

Now, we choose a reference power input of one watt, and we calculate both direct and reverberant levels at the back of the room (point A in Figure 13-3A):

$$\text{Direct field} = 118 - 20 \log (22) = 91 \text{ dB-SPL}.$$

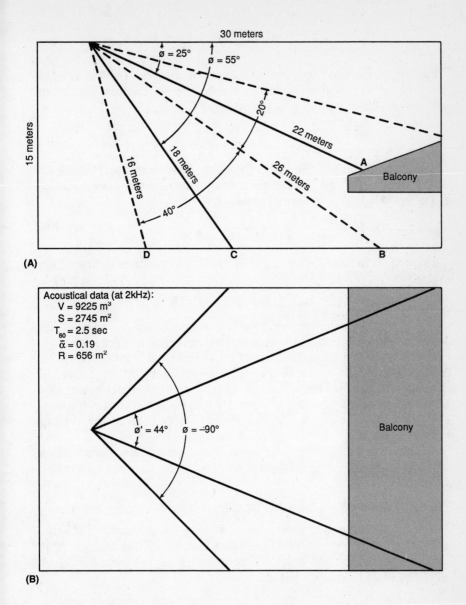

30 meters

ø = 25°

ø = 55°

20°

22 meters

16 meters

18 meters

26 meters

15 meters

A

Balcony

40°

D   C   B

(A)

Acoustical data (at 2kHz):
V = 9225 m³
S = 2745 m²
T_{60} = 2.5 sec
$\bar{\alpha}$ = 0.19
R = 656 m²

ø' = 44°   ø = –90°

Balcony

(B)

Figure 13-5. A design example using two HF horns. **(A)** Side elevation view. **(B)** Plan view.

The reverberant level can be calculated from equation 2-25:

$$L_{rev} = 10 \log (.18/656) + 126$$
$$= 90 \text{ dB-SPL}$$

The two values are very close, and this indicates that point A is just a little less than critical distance. As a check on our calculations, let us calculate critical distance using equation 2-24:

$$D_c = .14\sqrt{(45)\,(656)}$$
$$= 24 \text{ meters}$$

Had we not been provided with the value of average absorption, $\bar{\alpha}$, we could calculate it from the Norris-Eyring reverberation time equation (Equation 2-16) by rewriting it as follows:

$$\bar{\alpha} = 1 - \exp \left(\frac{-.16\,V}{ST} \right) \qquad \text{13-2}$$

At point B, the far throw horn, by itself, will produce a direct field level given by:

Level (direct) = 118 - 6 - 20 log (26) = 84 dB

The –6 dB term in this equation comes about because point B lies along the –6 dB zone of the horn's vertical coverage pattern.

Now let us move on to coverage of the middle and front of the space. The best choice here would be a 60-by-40 degree horn aimed downward so that it and the far-throw horn overlap along their –6 dB zones. As a starting point in establishing relative drive levels, let us power the near-throw horn so that it just matches the far throw horn at the overlap point B.

We do not know what power input to the near throw horn will be required to produce a direct field of 89 dB at point B, so we must work backwards:

Level (1 meter) = 84 + 6 + 20 log (26) = 118 dB

From data on the near throw horn, we note its sensitivity of 115 dB, one watt at one meter, and this indicates that we will have to power it with 2 watts (noting the 3 dB difference) in order to produce a direct field level of 84 dB at point B.

When both horns are powered accordingly, the resultant level at point B will be 90 dB, some 6 dB greater, assuming coherent summing of their response. Further, the near throw horn will produce direct field levels of 93 dB-SPL at point C and 88 dB-SPL at point D, as shown below:

Level at C (direct) = 118 - 20 log (18) = 93 dB
Level at D (direct) = 118 - 20 log (16) - 6 = 88 dB

Thus, the following direct sound field levels would be observed when the far-throw horn is powered by one watt and the near-throw horn with two watts:

Point A   91 dB-SPL
Point B   90
Point C   93
Point D   88

In final system checkout, the drive level to the near-throw horn might be lowered just a little in order to equalize levels in the house. This would be a matter of judgement on the part of the designer.

Having added the second horn, we must now make another reverberant field calculation. Keeping the input powers the same, we calculate, using equation 4-5, the efficiency of the near-throw horn. Its DI is 13 dB and its sensitivity is 115 dB, 1 watt at 1 meter. This indicates an efficiency of 20%. Since the power input is two watts, there will be 0.4 acoustic watts produced by the near-throw horn.

From our previous calculations using only the far-throw horn, we determined that that horn produced 0.18 acoustic watts when powered with one watt. Thus, the total acoustical power will be 0.58 acoustic watts.

Using equation 2-25 we now calculate the new reverberant level of 95 dB-SPL when both near- and far-throw horns are energized. Calculating the direct-to-reverberant ratios at the four points, we have:

Point A   −4 dB
Point B   −5 dB
Point C   −2 dB
Point D   −7 dB

Further points in the room, notably those off to the sides, could be calculated and compared with the reverberant level. We will leave it to the reader to do that.

INTELLIGIBILITY CALCULATIONS

At this point, the designer can refer to the data of Figure 12-4, assuming that the direct-to-reverberant calculations were made in the 1-2 kHz range, and an estimate of the system's intelligibility can be made. When this analysis is made, it is clear that most of the points analyzed in the room will yield acceptable, or better, articulation.

The room constant used in our calculations was the form defined in equation 2-23. However, if we assume that the bulk of the sound power radiated from the central array was aimed at the fairly absorptive audience area, then we might use the form of R defined in equation 2-26. More generally, not all of the power is directed at the audience; a better figure would be two-thirds incident on the audience and one-third in other directions. Therefore, we could define R′ as follows:

$$R' = S\bar{\alpha}/(.4 - \bar{\alpha}/3) \tag{13-2}$$

In this equation, we have assumed that the absorption coefficient of the audience is 0.95 in the 2 kHz band; therefore the equation should only be used in that frequency range for intelligibility estimates using the method of Peutz.

In this form, the equation assumes that the audience area is fully occupied, while equation 2-23 assumes that the house is empty. Both equations are useful, providing as they do worst and best cases for intelligibility estimates.

SYSTEM IMPLEMENTATION

Mechanical Considerations

Most manufacturers provide larger horns with 500-Hz cut-off frequencies as well as smaller ones with 800-1000 Hz cut-off. Further, the larger devices maintain their pattern control down to the 500-Hz range, while the smaller ones tend to lose control around 1 kHz.

For systems in very reverberant spaces, the additional pattern control in the 500-1000 Hz range is desirable, since it will maintain higher direct-to-reverberant ratios over a larger portion of the speech range. In spaces with shorter reverberation times, the smaller devices will work very well. There are also cost factors to be considered, and a carefully prepared specification will include acoustical analyses at 500 Hz and possibly 125 Hz as well.

Mechanical and Electrical Considerations

Figure 13-6 shows three possible physical realizations of the design we have discussed. At A, we show large horns for both far and near throw. A single 380 mm (15″) LF device has been added to complete the speech spectrum down to about 100 Hz. The realization shown at B uses a large horn for far throw and a smaller one for near throw, while the realization at C uses small components for both.

In Figure 13-6, we have indicated the relative sensitivities of the three devices. For single amplification, which is an acceptable alternative for a speech only system, the details of powering are shown in Figure 13-7A, while at B we show details of biamplification.

Maximum acoustical levels are reached when the total system power is increased up to the point at which the first element in the system reaches its power limit. In a system such as this, it is clear that the LF portion will overload first. With its sensitivity of 100 dB-SPL, one watt at one meter, and an assumed continuous power input rating of 150 watts, it is clear that maximum direct field levels of some 93 dB-SPL

HF, far-throw: Sensitivity = 118 dB (1w, 1m)
HF, near-throw: Sensitivity = 115 dB (1w, 1m)
LF: Sensitivity = 100 dB (1w, 1m)

(A)

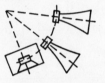

Same sensitivity as (A)

(B)

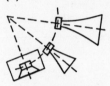

Same sensitivity as (A)

(C)

Figure 13-6. Mechanical aspects of system realization. **(A)** The use of large HF horns is recommended for larger, more reverberant spaces; note that voice coils are located in the same arc. **(B)** The use of large far-throw horns and a short near-throw horn; recommended in less reverberant spaces. **(C)** The use of small HF horns; this economical approach will not provide as good far-throw control in the 500-1,000 Hz range as either **(A)** or **(B)**.

will exist at a distance of 26 meters (point B). The actual level would be some 4 dB greater, inasmuch as the direct-to-reverberant ratio just below the balcony is –4 dB. The resulting level of 97 dB-SPL is certainly more than enough acoustical level for any speech requirement in a theater or house of worship.

GENERAL COMMENTS ON CENTRAL ARRAYS

An array should never have more components than necessary to do the intended job. Multiple components covering the same area are occasionally necessary for increased power output, but their specification should be carefully considered.

Splayed horns for increased angular coverage should, if at all possible, overlap along their –6-dB zones, and the high-frequency drivers should lie along the surface of a sphere, thus ensuring electrical coherence of their drive signals.

For systems which are not biamplified, it is imperative to determine the proper poling of high- and low-frequency sections of the system before it is rigged into place. It goes without saying that professional sound contractors and system installers will always set up an array in

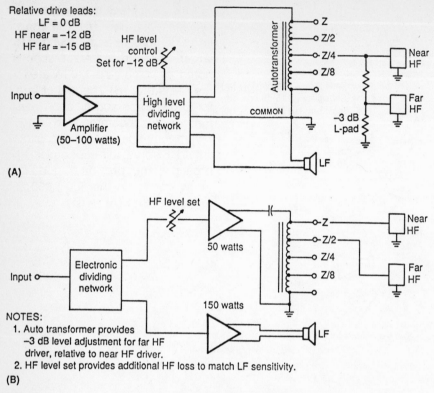

Figure 13-7. Electrical aspects of system realization. **(A)** The use of a high-level dividing network and a single power amplifier. **(B)** The use of bi-amplification.

their construction shop and check out all components before the materials are sent out on a job.

It is a rare indoor long-throw application today which cannot be best handled by a single large 20-by-40 degree constant coverage type horn. In years past, this application was often met by stacking four or more radial horns in a vertical line.

Recently, Keele[1] has developed a defined coverage horn that provides wide coverage toward the front of a room and narrow coverage toward the back. Figure 13-8A shows the specific rectangular aspect ratio for which the JBL model 4660 system is targeted, and at B we show the unique appearance of the system. Such a system as this, when used in a space having the desired aspect ratio, or similar ratios, can provide both near and far coverage without the expense of additional near-throw horns and their inevitable interference problems.

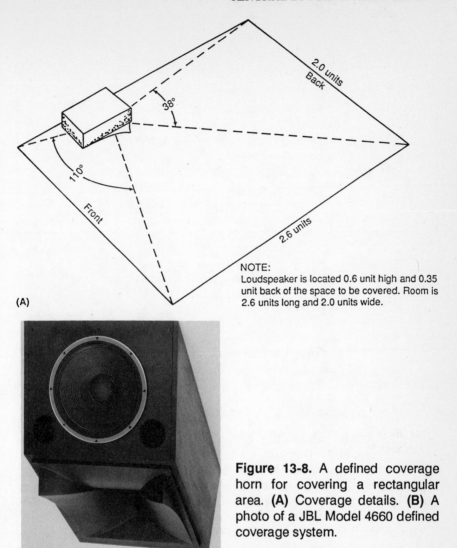

NOTE:
Loudspeaker is located 0.6 unit high and 0.35 unit back of the space to be covered. Room is 2.6 units long and 2.0 units wide.

(A)

(B)

Figure 13-8. A defined coverage horn for covering a rectangular area. **(A)** Coverage details. **(B)** A photo of a JBL Model 4660 defined coverage system.

Hybrid Systems: Implementation of Time Delay

Suppose that our estimates of intelligibility toward the rear of the room in the previous example indicated that there would be problems. Then, our best choice would be to limit the central array to coverage of, say, the front two-thirds of the room, implementing delayed loudspeakers to cover the rear one-third of the space. The rear loudspeakers would most likely have to be placed on the side walls, each aimed at one side of the rear audience area. Some details of this are given in Figure 13-9A and B.

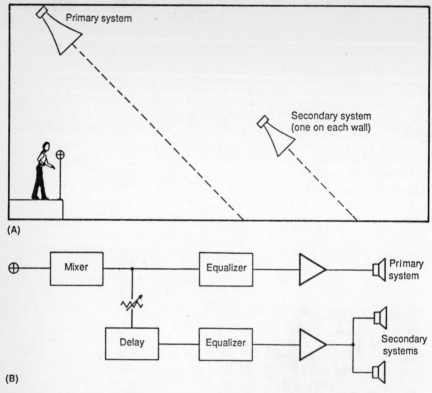

Figure 13-9. Hybrid systems. **(A)** Side elevation. **(B)** Schematic.

The rationale behind this design approach is that the delayed loudspeakers, since they are quite close to their intended listeners, can be operated at quite low power inputs, and thus not contribute substantially to the reverberant field of the room. Further, they are specifically aimed at the fairly absorptive audience, and reflected acoustical power will be quite low.

The setting of the time delay would be made using the principles discussed in Chapter 3. The delay time from the main array is calculated back to the point where the secondary loudspeakers are located, and an additional 10 milliseconds is added. This value of time delay should be implemented; the purpose of the added 10 milliseconds is to further ensure that the secondary channels can be raised in level, as may be needed, while maintaining localization toward the front central array.

The same approach will apply to under balcony seating locations in auditoriums and houses of worship, except that a number of small loudspeakers, usually 125 mm (5″) or 200 mm (8″), would be located in the balcony soffit rather than secondary loudspeakers on the side walls.

Some Caveats

In many houses of worship, it is impossible to locate a central array in the desired position. A satisfactory alternative is to locate a single array to one side, preferably the side where the pulpit is located. Better yet, a pair of arrays, one on each side, will provide greater realism, if each one is used only for speech originating on the appropriate side of the chancel.

It is strongly recommended that both arrays not be used at the same time. The reasons for this are aptly demonstrated in Figure 13-10. Interferences will color the sound considerably for many listeners, and delay effects, especially if the loudspeakers are more than about 12 meters (40′) apart will have adverse effects on system intelligibility.

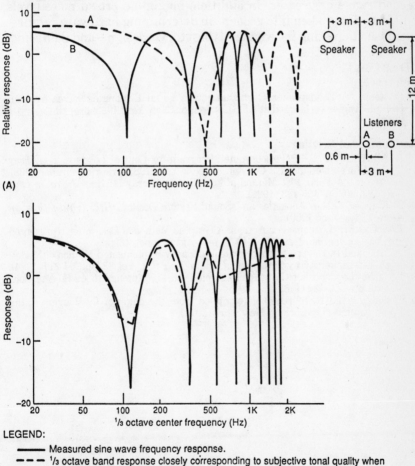

LEGEND:

─────── Measured sine wave frequency response.

– – – ⅓ octave band response closely corresponding to subjective tonal quality when listening to normal program material. Above 1 kHz subjective response is

(B) essentially flat.

Figure 13-10. Interference problems in split arrays (JBL Data).

We cannot overstate the requirement for good acoustical conditions for best operation of a central array. The expectations of these systems are often quite high, and they are often implemented at considerable expense. Most deleterious perhaps are reflections off the back wall, and their suppression can only be accomplished through appropriate acoustical treatment. Wherever possible, an acoustical consultant should be retained for advice in such matters.

Central arrays are often set up out of doors for covering large audience areas. As often as not, these are not permanent systems, and they are not always given the degree of fine tuning that we expect of the best indoor systems. While there is no reverberant field in the strict sense, there may be considerable discrete reflections which can adversely affect system intelligibility if they are in the 50 millisecond range or greater. In addition, maximum crowd noise levels will have to be taken into account in determining hardware requirements. Anticipating the worst, it is best to overdesign—and then some.

CHAPTER 13:

Reference:

1. D. Keele, "A Loudspeaker Horn that covers a Flat Rectangular Area from an Oblique Angle," (presented at AES Convention, New York 1983; preprint number 2052).

Recommended Reading:

1. D. Albertz, et al., "A Microcomputer Program for Central Loudspeaker Array Design," (presented at AES Convention, New York 1983; preprint number 2028).
2. F. Becker, "A Polar Plot Method of Loudspeaker Array Design," *J. Audio Eng. Soc.*, Vol. 30, pp. 425-433 (1982).
3. D. Klepper, "Room Acoustics and Sound System Design," *IRE Transactions on Audio*, (May-June 1960).
4. T. McCarthy, "Loudspeaker Arrays: A Graphic Method of Designing," (presented at AES Convention, New York 1978; preprint number 1398).
5. J. Prohs and D. Harris, "An Accurate and Easily Implemented Method of Modelling Loudspeaker Array Coverage," *J. Audio Eng. Soc.*, Vol. 32, pp. 204-217 (1984).
6. T. Uzzle, "Loudspeaker Coverage by Architectural Mapping," *J. Audio Eng. Soc.*, Vol. 30, pp. 412-424 (1982).
7. Various, Sound Reinforcement (compiled from the pages of the Journal of the Audio Engineering Society, New York, 1978).

CHAPTER 14

Distributed Systems for Speech Reinforcement

INTRODUCTION

In many rooms, typically large houses of worship with long reverberation times, a central array may not provide sufficient intelligibility. The reason is generally that the direct-to-reverberant ratio cannot be kept sufficiently high. Under these conditions, a distributed system with time delay zoning is indicated. The distributed loudspeakers are placed as close to the listeners as possible. Column loudspeakers located on the side walls are the most common solution. In large ballrooms, distributed systems are designed around ceiling loudspeakers of a fairly rugged sort, usually coaxial models of 300 mm (12″) or 380 mm (15″) diameter.

The proper implementation of time delay may be costly, but it is the only way that these systems can approach the natural effect that we take for granted in a properly designed central array.

IMPLEMENTATION OF TIME DELAY

As we discussed in Chapter 3, a reflection delayed up to 30 msec relative to a primary signal will not be heard as such, although it will contribute to the overall impression of loudness. In Figure 14-1, we show the effect of a delayed signal on a listener. For a listener who is close enough to a single loudspeaker in a distributed array, simply delaying the feed to that loudspeaker to an amount equal to the acoustical time delay from the source of sound will result in localization between the two sources when the two loudspeakers produce equal levels (as shown at A). If the signal to the distributed loudspeaker is further delayed some 10 to 20 msec, then the listener will clearly localize the sound at the source at the front of the room, even though the delayed sound may be increased up to 10 dB louder (as shown at B).

Thus, proper implementation of time delay involves calculating the acoustical delay from the source to the listener, and then adding a 10 to 20 msec margin to it. It is impractical to zone each loudspeaker with its own time delay; usually, it is sufficient to zone the listening area in increments of 20-25 msec. For a listener who is seated midway between two loudspeakers in different time zones, the effect may be one of some coloration of the sound, due to reinforcements and cancellations.

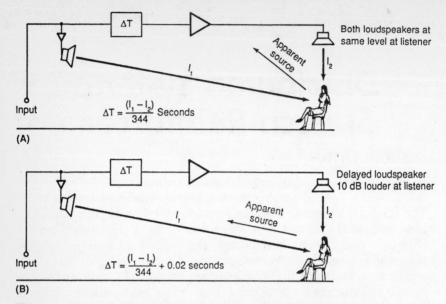

Figure 14-1. Implementation of time delay. **(A)** Delay equal to path length (length in meters). **(B)** Delay equal to path length plus 20 msec.

However, the predominance of the direct sound from the front of the room will probably obviate this.

For the best illusion of sound originating at the front of the room, a "target" loudspeaker located quite close to the talker is recommended. Such a loudspeaker can be located in the canopy above a pulpit, for example. The effect of this loudspeaker is to add focus to the sound originating from the talker. More often than not, the signal to that loudspeaker will be equalized so that high frequencies are subtly boosted.

SOME DESIGN EXAMPLES

Use of Sound Columns

Figure 14-2 shows a distributed system in a large cruciform church. Typically, such structures have reverberation times at mid frequencies as high as four or five seconds. Under such conditions, it is important that the distributed loudspeakers be located as close as possible to the listeners so that they may be driven at fairly low power inputs. In this way, the direct sound field components can be kept high relative to the reverberant field.

Use of Loudspeakers in Chandeliers

From a purely visual point of view, the location of loudspeakers in chandeliers is a desirable thing—provided that the coverage can be dense enough and the loudspeakers offer sufficient output capability.

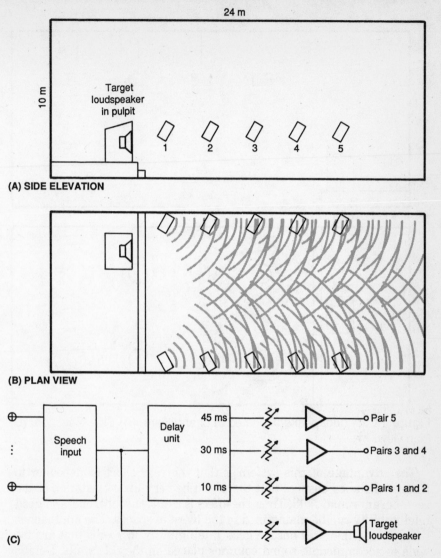

Figure 14-2. A distribution system in a house of worship. **(A)** Side elevation. **(B)** Plan view. **(C)** The electrical diagram.

An example is shown in Figure 14-3. More often than not, the requirements of lighting and those of sound reinforcement do not mesh too well.

A Pewback System

The ultimate distributed system in a house of worship is a pewback system. In such a system, small loudspeakers are located every two meters or so on the backs of the pews. They are zoned in time delay as required.

253

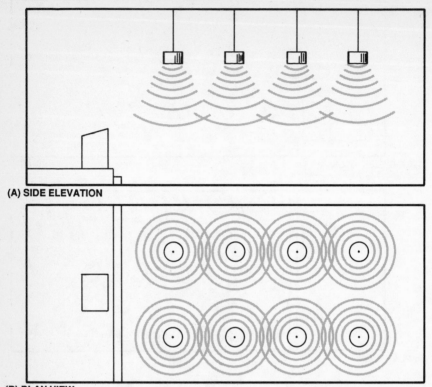

(A) SIDE ELEVATION

(B) PLAN VIEW

Figure 14-3. Loudspeakers located in chandeliers. **(A)** Side elevation. **(B)** Plan view.

The advantage of this system is that the very short loudspeaker to listener distance ensures that there will be very little excitation of the reverberant sound field. Thus, the effect is a quite natural one, and good intelligibility can be maintained in the livest of spaces. The approach is the only one that will ensure good intelligibility in rooms that are too wide to accommodate sound columns placed on the side walls. Details are shown in Figure 14-4.

SYSTEMS FOR BALLROOMS

General Comments

Ballrooms are rarely as reverberant as large houses of worship, and the specification of distributed systems in these spaces arises from a different set of requirements. Generally, the relatively low ceiling in such spaces indicates that a central array will not work well. However, the main advantage may come from the additional requirement of partitioning and rearranging the space for different functions, in which

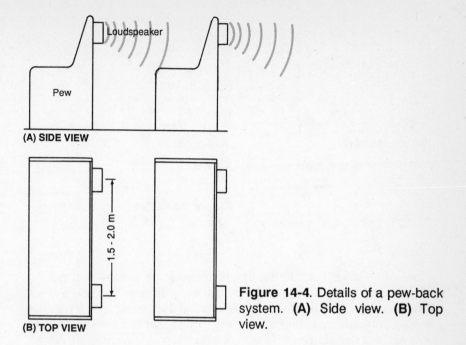

Figure 14-4. Details of a pew-back system. **(A)** Side view. **(B)** Top view.

case the zoned loudspeakers will allow a high degree of flexibility.

A straightforward layout for a large ballroom is shown in Figure 14-5. For the space shown, quite rugged coaxial loudspeakers should be located in a triangular, or criss-cross, array with spacing close enough so that the overlap in the 2 kHz range is as indicated in Figure 14-6. An alternate array is shown in Figure 14-7. While the array of Figure 14-6 is preferred for most uniform coverage, both approaches will work well, and wide-band variations at the ear plane will be no more than plus or minus 2 to 3 dB.

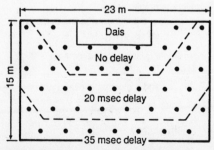

Time delay with a distributed ceiling system.

Figure 14-5. A ballroom system.

255

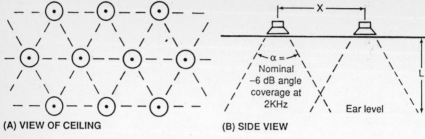

(A) VIEW OF CEILING   **(B) SIDE VIEW**

Equations for determining coverage requirements: $X = 2L\tan(^\alpha/_2)$

Approximate quantity of loudspeakers $= \dfrac{\text{Area}}{0.8X^2}$

NOTE:
Both area and X must be in consistent units.

Figure 14-6. Details of the triangular array.

More often than not, the luxury of these dense arrays will not be affordable, and the consultant or contractor will have to work around the requirements for lighting and ventilation duct work. For irregular arrays, it is best to calculate direct field coverage at ear level at selected points making use of inverse square relationships and combining power levels using Figure 2-8.

Multi-purpose Spaces

In most new hotel and convention complexes, movable wall sections make possible the structuring of the space for a wide variety of applications. It is essential that the ceiling loudspeakers be grouped corresponding to the smallest area the space can be subdivided into. In this manner, the loudspeakers can always be zoned for the function at hand. A separate amplifier will of course be required for each group. Details of this are shown in Figure 14-8.

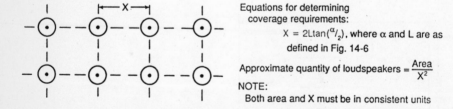

Equations for determining
coverage requirements:

$X = 2L\tan(^\alpha/_2)$, where α and L are as
defined in Fig. 14-6

Approximate quantity of loudspeakers $= \dfrac{\text{Area}}{X^2}$

NOTE:
Both area and X must be in consistent units

Figure 14-7. Details of the square array, ceiling view.

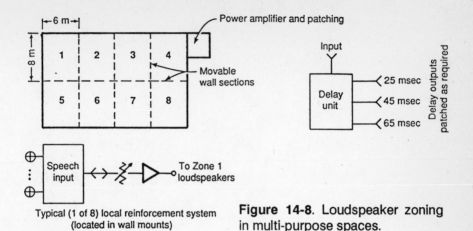

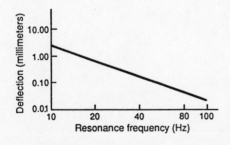

Typical (1 of 8) local reinforcement system
(located in wall mounts)

Figure 14-8. Loudspeaker zoning in multi-purpose spaces.

HARDWARE REQUIREMENTS

No less than in the case of central arrays, a distributed system places great demands on hardware. Only the most rugged amplifiers, patching facilities, and loudspeakers should be specified. In particular, patching and switching gear should be chosen for the heavy duty it is to see.

For speech requirements, response below 80 Hz is not required, and this consideration will lead naturally to loudspeaker enclosures of reasonable size. In all cases, manufacturers' recommendations regarding enclosure details should be followed.

Figure 14-9. Cone deflection due to gravity.

The sag of loudspeaker cones due to gravity will, if it is excessive, increase distortion at high drive power inputs. The data of Figure 14-9 indicates the degree of this effect. Obviously, where possible, loudspeakers with high free air resonances should be specified.

Paging Systems

INTRODUCTION

Paging systems are almost invariably designed as distributed systems. However, unlike those systems discussed in Chapter 14, there is no attempt to focus the source of sound toward a given talker. Paging systems are specified for most large public areas, such as transportation terminals, stores, and some office spaces, where messages must be transmitted either to groups or to individual persons.

Since most public areas have a ground noise level which can range from moderate to quite high, an effective paging system must have power reserve to be clearly audible above the highest noise levels. It is therefore essential that an analysis of system intelligibility be made before the system is specified. The Articulation Index (AI) method of Kryter[1] is generally the preferred method of estimating the intelligibility of paging systems, since that method takes into account the effects of the noise spectrum over the range from 250 Hz to 4 kHz.

We will begin with a study of constant voltage power transmission, since this system is invariably used in paging applications.

CONSTANT VOLTAGE TRANSMISSION SYSTEMS

In the normal amplifier-loudspeaker interface, we are accustomed to viewing the amplifier as delivering its full output power into a specific load impedance. For example, an amplifier may be designed to deliver its full power output of, say, 200 watts into an 8-ohm load. We then know that when we place an 8-ohm load impedance across the output, the amplifier will deliver rated power when it is driven to its rated output voltage.

However, when the effective load on the amplifier consists of many smaller loads in parallel, as in the case of a distributed system, an accounting of the actual load impedance can be quite cumbersome, as indicated by equation 1-10. The constant voltage system of power distribution was developed to simplify the accounting of loads across an amplifier.

In a constant voltage system, the full output power of an amplifier, whatever its power rating, is available at a fixed voltage. For most paging systems in the United States, the output is 70 volts RMS. Other standards less commonly encountered are 100 volts RMS (in Europe) and 25 volts RMS.

In such a system, line distribution transformers are placed across the 70-volt line. Primary taps are indicated in watts drawn from the line, and the secondary taps may be indicated in nominal loudspeaker impedances of 4, 8, or 16 ohms. The designer of the system merely has to count watts drawn from the line by the various primary power settings. If a 200-watt amplifier is used, then the sum of all the powers drawn by the transformer-loudspeaker combinations should not exceed 200 watts. Under this condition, the amplifier will be properly loaded. Details of this are shown in Figure 15-1.

Line and Transformer Losses

Since most paging systems are subject to stringent budget considerations, fairly low-cost components are generally used. The transformers in particular are quite variable, and their performance at low frequencies should be accurately known. Through the use of high-pass filtering, say, at 100 Hz, many of the LF problems can be obviated.

The insertion loss of a distribution transformer should be carefully noted. For example, if the insertion loss of a given transformer is 1 dB, then when its 5-watt tap is placed across the 70-volt line, it will deliver 4 watts to the load.

The power tap on the transformer should represent the actual power drawn from the line, *NOT* the actual power delivered to the loudspeaker. Make absolutely sure that the transformers you intend to use follow this convention. If this should not be the case, then there is a strong risk of overloading the amplifier. Details of these considerations are shown in Figure 15-2.

Line losses are generally small in 70-volt systems inasmuch as the individual load impedances are quite large in comparison with the resistance of the wire used in connecting them. Computing the exact

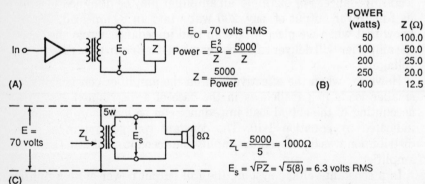

POWER (watts)	Z (Ω)
50	100.0
100	50.0
200	25.0
250	20.0
400	12.5

$E_O = 70$ volts RMS

$\text{Power} = \dfrac{E_O^2}{Z} = \dfrac{5000}{Z}$

$Z = \dfrac{5000}{\text{Power}}$

$Z_L = \dfrac{5000}{5} = 1000\Omega$

$E_s = \sqrt{PZ} = \sqrt{5(8)} = 6.3$ volts RMS

Figure 15-1. A 70-volt constant voltage distribution system. **(A)** The basic 70-volt system; Z=total load across the line. **(B)** Rated load impedances for various 70-volt power amplifiers. **(C)** Calculating the impedance (Z_L) of a given wattage tap across a 70-volt line.

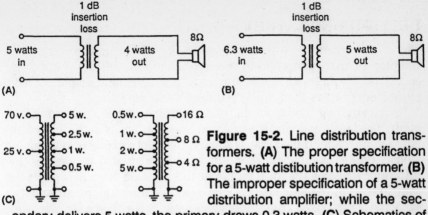

Figure 15-2. Line distribution transformers. **(A)** The proper specification for a 5-watt distibution transformer. **(B)** The improper specification of a 5-watt distribution amplifier; while the secondary delivers 5 watts, the primary draws 0.3 watts. **(C)** Schematics of typical distibution transformers; the transformer on the left is designed to work into an 8-ohm load, the transformer on the right is designed to work only across a 70-volt line.

losses in a typical distributed system is cumbersome, due to the many loops in the circuit. Most designers make a single, worst case calculation, and this assumes that all loads are seen in parallel at the end of the entire wire run. Details of this are shown in Figure 15-3. A general design criterion is that line losses, calculated in this manner, should not exceed 0.5 dB.

The loss indicated by the equation in the figure represents the difference between the power actually delivered to the load as compared with the power intended to be delivered to the load. It takes into account both line losses and a loss due to the impedance mismatch caused by the added wire resistance.

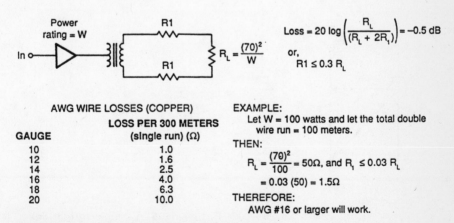

Figure 15-3. Line losses in 70-volt systems.

DETERMINING LAYOUT DENSITY

The rules regarding pattern overlap in the 2 kHz range discussed in Chapter 14 are appropriate in this application as well. However, labor cost is usually an important factor, and a paging system should not have more loudspeakers than really required to do the job.

In determining the layout density, a number of calculations must be made. The level variation as a function of array density is the first calculation to be made. If the array is square, the level directly below a single loudspeaker is calculated. Then, the level is calculated for the position midway between a diagonal pair. For this position, we will observe the contribution of the four nearest loudspeakers, ignoring, for simplicity's sake, the contribution from those loudspeakers which are farther away. Note from Figure 15-4 that these calculations can be normalized to the reference level, L_1, of the single loudspeaker directly overhead. In making these calculations, we determine the effect of a single loudspeaker at the position midway between four loudspeakers. The level at this position, L_2, will be modified by the loudspeaker's pattern loss at the particular off-axis angle, *beta*, and the resultant level will be the combination of powers produced by the four loudspeakers, or 6 dB greater than produced by a single loudspeaker. We are assuming that all loudspeakers are driven with the same power input. For values of

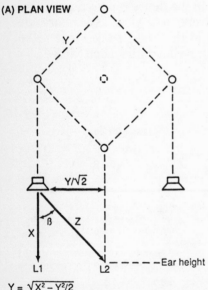

(A) PLAN VIEW

$Y = \sqrt{X^2 - Y^2/2}$

$\beta = \text{arc cos } (^X/z)$

$L2 = L1 - 20 \log (^Z/x) - \text{Pattern loss @}\beta + 6 \text{ dB}$

(B) SIDE VIEW

Figure 15-4. A square loudspeaker array.

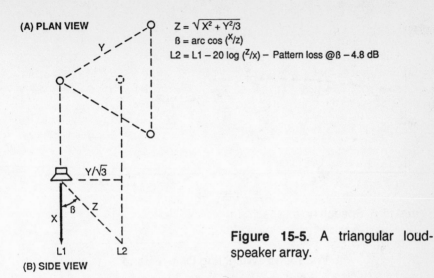

(A) PLAN VIEW

$Z = \sqrt{X^2 + Y^2/3}$

$\beta = \arccos{(X/z)}$

$L2 = L1 - 20 \log{(Z/x)} - \text{Pattern loss } @\beta - 4.8 \text{ dB}$

$Y/\sqrt{3}$

L1   L2

(B) SIDE VIEW

Figure 15-5. A triangular loud-speaker array.

Y which are at least twice X, this procedure will yield level calculations within one dB of the actual levels.

For the triangular array, the calculations are as shown in Figure 15-5. Here, the level at the point midway between three loudspeakers will be augmented by 4.8 dB, the level sum of the three equal powers.

A DESIGN EXAMPLE

Let us design a paging system for a transportation terminal. Reverberation is not assumed to be a problem in this particular space, but the noise level can be fairly high during peak travel periods. The noise spectrum is assumed to be essentially a speech spectrum, and the maximum RMS noise level, as measured on the flat scale of a sound level meter, is 65 dB-SPL. We will determine the intelligibility estimate of the system by using the Articulation Index (AI) method.

Our goal is to select components for a square ceiling array that will satisfy the needs for good intelligibility under the noisiest conditions. The analysis will be a detailed one, and we will make it in several steps.

The Noise Spectrum

A typical speech spectrum is shown in Figure 15-6. The values given for each of five octave bands represent the levels in those bands below a given wide-band noise level. In our example, the wide-band noise level is 65 dB-SPL; thus, the levels in each octave band will be as indicated in Table 15-1.

In some instances the noise spectrum in a space will not necessarily be a speech spectrum. In such cases, an acoustical consultant should provide the actual maximum measured noise levels in each band.

263

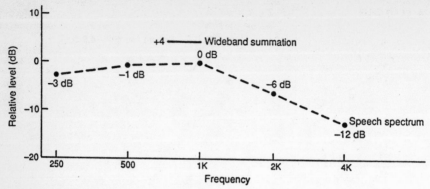

Figure 15-6. Speech noise spectrum.

	Wideband Level	Weighting	Octave Band Levels
250 Hz Band	65	−7	= 58 dB-SPL
500 Hz "	65	−5	= 60 "
1kHz "	65	−4	= 61 "
2kHz "	65	−10	= 55 "
4kHz "	65	−16	= 49 "

Table 15-1. Relating wideband noise level to octave band levels.

Loudspeaker Characteristics

For this design example, we will use a typical 200 mm (8″) two-way (coaxial) design with an integral line-to-voice coil transformer which can be tapped at 5, 2.5, 1, 0.5, or 0.25 watts. The nominal sensitivity of the device is 97 dB, one watt at one meter, and its on-axis response is flat over the frequency range we are interested in. The directional characteristics of this loudspeaker are shown in tabular form in Figure 15-7.

Room Dimensions

The space under consideration has the following dimensions:

Length = 50 meters
Width = 30 meters
Height = 6 meters

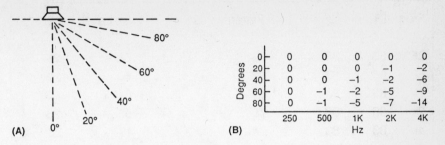

Figure 15-7. Directional characteristics of a 200 mm (8-in.) loudspeaker.

We will assume that average ear level for a seated person is one meter above the floor.

Trial Loudspeaker Layout

To begin with, we will assume a distance between loudspeakers 2.5 times the loudspeaker-to-ear distance, as shown in Figure 15-8. X, Y, and Z dimensions are as shown in the figure, as is the angle *beta*.

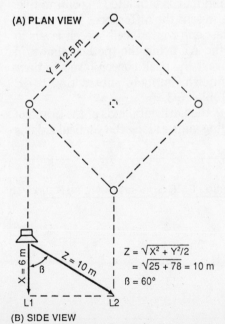

$$Z = \sqrt{X^2 + Y^2/2}$$
$$= \sqrt{25 + 78} = 10 \text{ m}$$
$$\beta = 60°$$

Figure 15-8. A trial layout.

Freq	L₁	L₂	L₂ & Pattern Loss	Noise Level	Δ
250	83	83	83	58	25
500	83	83	82	60	22
1k	83	83	81	61	20
2k	83	83	78	55	23
4k	83	83	74	49	25

Table 15-2. Determining peak speech-to-noise ratios

To test this assumed square layout, we will arbitrarily power the loudspeakers with one watt each and complete the data shown in Table 15-2. With its one-watt, one-meter sensitivity of 97 dB, a single loud-speaker will produce a level, L_1, of 83 dB-SPL directly beneath the loudspeaker at a distance of five meters. This is shown in column one.

The L_2 levels are tabulated in the second column as simply the L_1 levels minus six dB, the inverse square difference between X and Z plus the power summation of the four nearest loudspeakers, or +6 dB. Then, we account for the directional pattern loss at angle beta, as shown in column three.

The noise level in each band is tabulated in column four, and the difference between columns three and four is tabulated in column five.

These final values in the table represent the differences between the maximum RMS noise levels and the peak reproduced speech levels in each band. We can now calculate the AI, using the method shown in Figure 12-1, but first we must convert the peak speech levels to their effective RMS levels. This we accomplish simply by subtracting 12 dB from the peak levels, as shown in Table 15-3.

We can now enter these values into the horizontal axis of the graph of Figure 12-1 and sum the corresponding values along the vertical axis, as shown in Table 15-4.

Freq	Δ	RMS-to-Peak Ratio	RMS Speech-to-Noise Ratio
250	25	−12	13
500	22	−12	10
1k	20	−12	8
2k	23	−12	11
4k	25	−12	13

Table 15-3. Determining RMS speech-to-noise ratios.

Freq	AI Calculation
250	0.05
500	0.11
1k	0.15
2k	0.25
4k	0.215
TOTAL	0.775

Table 15-4. Articulation Index (AI) calculation.

The resulting AI is 0.775. Referring to Figure 12-3, we can see that this AI corresponds to about 92% random syllable recognition, and the system appears excellent. By inspection, we determine that about twelve loudspeakers would be sufficient for this system. Considerations of electrical headroom would dictate that we tap the loudspeakers at least 6 dB above the speech peak power input level of one watt. Thus the 5-watt tap would be used, and a 120-to-150-watt amplifier would be indicated.

The foregoing analysis has been lengthy but not difficult. The method is essentially one of "cut and try." By computer implementation, it could be made into an interactive program that could optimize any aspect of the design.

Had the first trial not worked out favorably, the approach would then be to use a higher reference power to the loudspeakers. Failing that, then a denser array of loudspeakers would be required. The iterative nature of the design procedure is obvious.

EFFECTS OF BOTH NOISE AND REVERBERATION

The presence of reverberation in addition to noise further deteriorates the AI. While the precise quantitative effect of short reverberation times has not been clearly established in noisy environments, it is clear that longer reverberation times behave simply as added noise. To account for these, the levels of reverberation are calculated in each band and are added to the individual noise levels using the method shown in Figure 2-8.

Note that reverberant levels follow the speech levels, so that when the overall system output level is increased, the reverberant level may tend to dominate the fixed noise levels.

It is strongly recommended that the detailed calculations of AI in environments which are both noisy and reverberant be referred to experienced electroacoustical consultants who have had experience in this kind of system design.

Figure 15-9. Typical loudspeakers used for paging systems (JBL photo).

POWER CLASS OF COMPONENTS

Most indoor paging systems in terminals and office areas will make use of the type of componentry shown in Figure 15-9. The figure shows two sizes of loudspeakers, the smaller 125 mm (5″) model is especially useful for speech privacy systems. The decorative grille fits unobtrusively into most ceiling tile elements. The line-to-voice coil transformer fits conveniently onto the frame of the larger loudspeakers.

For paging systems in quite noisy environments such as factories and warehouses, devices called paging horns are generally used. These are small integral horn-compression driver units with sensitivities of the order of 100-105 dB, one watt at one meter. Often, they have power input capability of 20 to 30 watts. Many models contain integral matching transformers and high-pass filtering. The devices as a class are inherently band-limited at low frequencies, cutting off abruptly below about 100-125 Hz.

In certain industrial applications, these devices must be explosion proof, due to the presence of flammable volatiles in the air. All electrical components used in such environments are carefully regulated by prevailing fire codes.

CHAPTER 15:
1. K. Kryter, "Methods for the Calculation and Use of the Articulation Index," *J. Acoustical Soc. Am.*, Vol. 34, p. 1689 (1962).

Artificial Ambience Systems

INTRODUCTION

Electroacoustical modification of the properties of performance spaces has been practiced to a limited degree for several decades. For the most part, European organizations, most notably, Philips in Holland, have led the way in research in this area.

In this country, Paul Veneklasen was the pioneer, with his work in auditorium synthesis dating from the sixties. In more recent times, Christopher Jaffe has designed complex electroacoustical ambience systems for music performance.

In many quarters, there still remain prejudices against such systems. To a large extent, the criticisms are valid ones, since many systems have been improperly specified and maintained.

The nagging problems have been the following:

1. Insufficient specification. Too often, there have not been enough channels to create a truly natural effect.

2. Reliability. Natural acoustics will always work—and that is the logical comparison to be made with an electroacoustical system. A system made up of many channels becomes, statistically, a failure-prone one.

3. Not tamper-proof. It seems that there is always a way to get at the controls of the system, and there is usually somebody who thinks he can set the system up better than the design consultant did.

Today, there is more need than ever before for electroacoustical ambience systems, inasmuch as multipurpose halls are becoming more useful. A typical concert hall, for example, cannot hold more than about 2700-3000 patrons without serious acoustical problems related to proper listening of classical music. Proper implementation of an electroacoustical ambience system holds the promise for increased seating capacity with satisfactory acoustics at a lower cost per patron. Alternatively, the demands for lectures or motion picture presentations in typical spaces call for lower reverberation time than for music. An electroacoustical ambience system can be tailored for each event as needed.

There are, broadly speaking, three methods employed here: *sound field amplification*, as developed by Philips, *sound field modeling*, or auditorium synthesis, and *assisted resonance*, as developed for Festival Hall in London.

SOUND FIELD AMPLIFICATION

In this method, many individual microphone-amplifier-loudspeaker channels are located on the wall and ceiling boundaries of a performance space. There may be as many as 100 or so of these channels, and each one is operated at fairly low gain. Sound impinging on a microphone is reradiated at a level corresponding to a reflection from a surface *less absorptive* than the actual boundary. In this way, the room becomes more live and reverberant. Let us work out an example.

Assume we have a space with the following characteristics: L = 50m; W = 30m; H = 15m and T_{60} = 1.25 sec.

Working back from the Eyring reverberation time equation, we can solve for the *average absorption coefficient, $\bar{\alpha}$.*

$$\bar{\alpha} = 1-\exp(-.16V/ST)$$

In this equation, V is the room volume in meters³, S is the total surface area in meters², and T is the reverberation time. Entering the values yields $\bar{\alpha}$ = 0.4 for the room.

We can now calculate the room constant, R, as follows:

$$R = S\bar{\alpha}/(1-\bar{\alpha})$$

Solving this equation yields R = 3600 meters².

Let us now calculate the reverberant sound pressure level in this room if a sound source is delivering 25 acoustical watts, the peak acoustical power output of a symphony orchestra[1]:

$$L_{rev} = 126 + 10 \log (W/R)$$

In this equation, W is the acoustical power in watts. Solving: L_{rev} = 126 + 10 log (25/3600) = 104 dB-SPL.

Thus, a symphony orchestra playing in this room could produce peaks in the reverberant field of 104 dB. However, the reverberation time in the room is only 1.25 seconds; ideally, for symphonic music, it should be about twice that amount, or 2.5 seconds.

Let us assume that we have a room of the same dimensions with a reverberation time of 2.5 seconds. Let us solve for the new value of $\bar{\alpha}$ which characterizes this new room:

$$\bar{\alpha} = 1-\exp (-.16V/ST) = 0.23.$$

The new room constant is 1613 meters², and we now calculate the new reverberant level in the room produced by 25 acoustical watts:

$$L_{rev} = 126 + 10 \log (25/1613) = 108 \text{ dB-SPL}$$

This new level is 4 dB greater than that observed in the original room, and if we are to simulate the effect electronically we will have to add to the room, via loudspeakers, peak-power capability some 4 dB greater than 25 watts:

Power$_{total}$ = 25 $(10^{4/10})$ = 63 watts.

Since 63–25 = 38 watts, we must add to the reverberant field a total of 38 acoustical watts when the normal acoustical level peaks out at 25 watts. If we split the load into 100 channels, each channel has only to deliver 0.38, or approximately 0.4 acoustical watts on peaks. Each of the 100 channels would have its level adjusted as shown in Figure 16-1. Both microphones and loudspeakers are located at wall or ceiling boundaries. When the microphone is placed in a 104 dB sound field, the electrical gain in the channel is adjusted so that the loudspeaker produces a free-field level of 108 dB at 1 meter. In so doing, it is radiating 0.4 watts, and the entire ensemble of 100 channels will radiate 40 acoustical watts.

The gain of this system is 4 dB; that is, with the system turned on, the reverberant sound field increases by 4 dB. In the process, the reverberation time has been increased by a factor of 2. Philips points out that the maximum gain such a system can handle without adverse effects is given by:

Gain$_{max}$ = $(n +50)/50$,

where n is the number of channels.

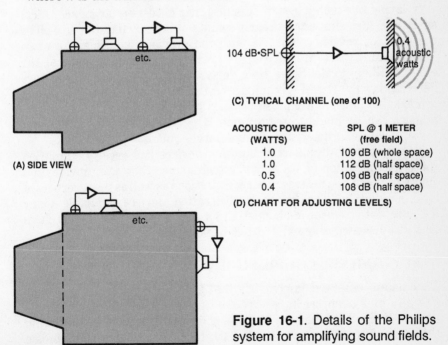

(A) SIDE VIEW

(B) PLAN VIEW

(C) TYPICAL CHANNEL (one of 100)

104 dB·SPL → 0.4 acoustic watts

ACOUSTIC POWER (WATTS)	SPL @ 1 METER (free field)
1.0	109 dB (whole space)
1.0	112 dB (half space)
0.5	109 dB (half space)
0.4	108 dB (half space)

(D) CHART FOR ADJUSTING LEVELS

Figure 16-1. Details of the Philips system for amplifying sound fields.

The *greater* the number of channels, the *lower* the gain at which each one has to work, and the less the tendency for the aggregate system to ring, or resonate, at particular frequencies. In short, the room and system will behave more like an acoustically live space. Of course, there must be adequate electrical power to drive each channel to the maximum level expected of it, depending on the kind of musical activity to be performed in the room.

Philips also points out that as little direct sound as possible from the stage should enter the microphones, since the intent of the approach is only to amplify the diffuse reverberant field of the room.

SOUND FIELD MODELING

In the preceding example, electroacoustics was employed to liven a room, but not to make it seem larger than it really is. With sound-field modeling, a fairly small, acoustically dead space may be transformed into a room much larger as well as more reverberant. Figure 16-2 shows how this may be accomplished. Views A and B show side and plan views of an auditorium, while C shows the electronic signal flow diagram. Stage microphones pick up the sound to be processed, and it is important that these microphone inputs be well isolated from the amplified and processed sounds in the house. Acoustical feedback can be a problem if care is not taken here.

In laying out such a system, the designer chooses a target acoustical space and then simulates its early sound field characteristics as well as the onset of reverberation. As with the previous example, the more loudspeakers there are, driven at low levels, the more natural the effect is likely to be. Since time delays, which are the essential cues determining the size of a room, are in the hands of the designer, some remarkable illusions are possible using such a system as this. Typically, a space can be made to seem much larger than it really is by using initial delays characteristic of much larger rooms.

The advent of digital reverberation devices has greatly simplified the implementation of systems like these, since they can provide, in a single package, the necessary early reflections as well as the reverberant field simulation. The better reverberation devices available today accommodate a stereo (2-channel) input and provide four output channels, two forward-oriented and two back-oriented.

STAGE-TO-HALL COUPLING

Veneklasen[1] observed fairly early that the upper portion of the stage house in many concert halls was often far more reverberant than the hall itself. He developed the method, shown in Figure 16-3, for getting the trapped sound out into the hall itself. This expedient represents a

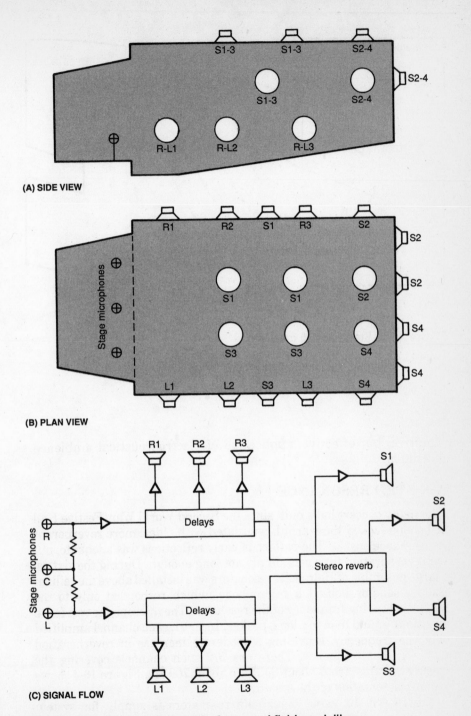

(A) SIDE VIEW

(B) PLAN VIEW

(C) SIGNAL FLOW

Figure 16-2. Details of a system for sound field modelling.

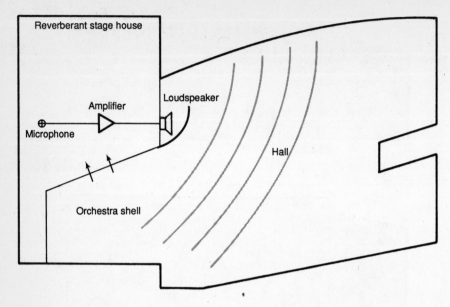

Figure 16-3. Stage-to-hall coupling.

primitive, but effective, application of electroacoustical ambience generation.

ASSISTED RESONANCE

Like many concert halls built since the Second World War, Festive Hall in London lacked the warmth associated with older, more reverberant performance spaces. The pattern of early reflections was adequate, but the reverberation time was simply not long enough. During the sixties, a large ensemble of Helmholtz resonators was installed above the ceiling. Each resonator housed a microphone, which responded only to the individual tuning frequency of the resonator. The microphone was fed to an amplifier and then to a loudspeaker. In short, each channel amplified only one frequency, providing a moderate increase in reverberation time for that frequency. There are 172 such channels covering the frequency range from about 60 Hz to about 700 Hz. Figure 16-4 shows the implementation of the system[2].

The reason for using Helmholtz resonators is simply for system stability and ensuring that the channels will not interact with each other.

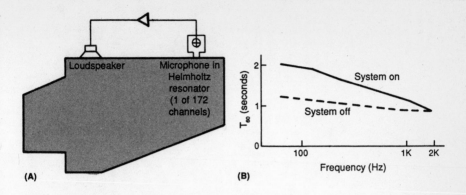

Figure 16-4. Assisted resonance.

CONCLUSIONS

We have seen how electroacoustical techniques can be used to simulate natural reverberant fields. In the future, we will probably see more such applications as the cost of digital signal processing continues to drop and as more engineers and architects are called upon to design more flexible performance spaces.

CHAPTER 16:
Reference:
1. P. Veneklasen, "Design Considerations from the viewpoint of the Professional Consultant," *Auditorium Acoustics*, pp. 21-42, Applied Science Publishers, London (1975).
Additional Reading:
1. Philips Product Bulletin: Multi-channel Reverberation System (published by Audio-video Systems Group).
2. P. Parkin, "Assisted Resonance," *Auditorium Acoustics*, pp. 169-179, Applied Science Publishers, London (1975).
3. Anon, "Sound System Design for the Eugene Performing Arts Center, Oregon," *Recording Engineer/Producer*, pp. 86-93 (December 1982).

CHAPTER 17

Speech Privacy and Noise Masking Systems

INTRODUCTION

Although they are not, strictly speaking, sound reinforcement systems, speech privacy and noise masking systems often make use of the same distributed loudspeaker array that may be used for paging systems. Thus, it is appropriate to discuss these systems in this handbook.

We have all had the experience of trying to sleep in an extremely quiet environment interrupted from time to time by discrete noises, such as a dog barking or automobile horn. The difficulty in falling to sleep under these circumstances arises from the discrete nature of the interfering noises against the extremely low background noise level.

Now, in contrast, imagine the background noise level raised by the gentle fall of rain on the roof or the rushing of a nearby brook. The new noise level is not high enough in itself to be disturbing, but it is high enough to mask the discrete sounds of dogs barking and horns honking. Thus, we are able to enjoy a pleasant night's sleep.

This is in essence how a speech privacy and noise masking system works. Such a system in effect reduces the signal-to-noise ratio of an undesired signal to a background noise level to zero or below. The need for these systems comes about through the implementation of open plan office spaces in many modern commercial buildings.

System Implementation

The basic layout is shown in Figure 17-1. A person seated at position B does not want to be disturbed by a talker at position A. In order to accomplish this, a partition is installed between the adjacent work areas, and a controlled masking noise is introduced into the area by way of a dense overhead array of loudspeakers.

Sound from A reaches the listener by way of diffraction around the partition as well as reflection from the ceiling. The sound arriving at B via these two paths is attenuated, especially at high frequencies, and it is low enough that it can be masked by a suitably equalized random noise signal introduced by the loudspeaker array in the ceiling. However, if the sound originating at A is too loud, the noise level that would be required to mask it would itself become a source of disturbance to the person seated at B. Conversely, if the attenuation from A to B takes place over a sufficiently large distance, and if the speech originating at A is of normal level, the partition may be eliminated.

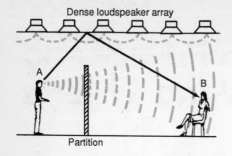

Figure 17-1. A speech-privacy, noise-masking system.

The nature of the masking noise itself is critical, both as regards spectrum shape and level. Figure 17-2 shows the family of Noise Criteria curves used for assessing noise levels in architectural spaces. They have very nearly the same shape as the familiar equal loudness contours which are illustrated in Chapter 3 (Figure 3-1).

Most people will be unaware of a noise spectrum corresponding to NC 30. While a noise spectrum corresponding to NC 35 will be apparent to some people, most people can adjust to it. NC 40, however, is the maximum noise which can be used in practice. The actual shape of the masking noise spectrum is often modified at both high and low frequencies to resemble the contour shown in Figure 17-3.

Having established an allowable masking noise spectrum and level which can be introduced into the space, the designer then turns his attention to the sound attenuation between adjacent work areas.

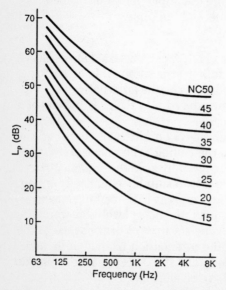

Figure 17-2. A family of noise criteria (NC) curves.

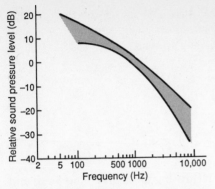

Figure 17-3. Octave band spectrum for random background noise, for speech privacy.

We now examine the curves shown in Figure 17-4. These are known as Sound Transmission Class (STC) curves, although in this application we may refer to them as Noise Isolation Class (NIC) curves. Fortunately, most acoustical barriers are more effective at high frequencies than they are at low frequencies, so that their transmission loss with respect to frequency is roughly the inverse of the equal loudness contours. A barrier or transmission path rated at STC 20 will introduce a loss as indicated by the corresponding curve.

STC 20 represents the minimum loss, referred to one meter, that should exist between A and B for successful masking, taking into account all of the acoustical paths between A and B. This consideration usually requires that the distance between A and B be no less than 3 to 4 meters (10 to 13 feet) and that partitions with a transmission loss between 400 and 2000 Hz of not less than 10 dB be used. The surfaces of the partitions should have a fairly high absorption coefficient, and they should extend to the floor. The partition height is a design variable that may be dictated by the degree of privacy required.

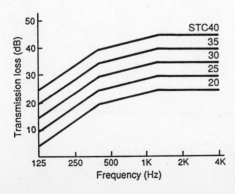

Figure 17-4. A family of sound transmission class (STC) curves.

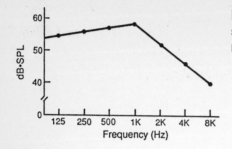

Figure 17-5. Long-term average spectrum at one meter for normal male speech.

It is generally felt that the sum of NC rating of the masking noise and the STC rating of the transmission path be equal to 60 or greater. In making this determination, a sound source at A is measured at a distance of one meter at one-third octave intervals over the range of 400 to 2000 Hz. Then the same measurements are made at B. The differences represent the interzone attenuation between A and B. The specific noise masking spectrum may then be shaped so that, on a one-third octave basis, the sum of the STC ratings and the NC ratings is equal to 60 or greater.

System Analysis

An analysis of the system proceeds as follows. Figure 17-5 shows the long-term average spectrum at a distance of one meter for normal male speech. Note that maximum levels are in the 500 and 1000 Hz bands and that they are very nearly 60 dB-SPL. We now show in Figure 17-6 the effect of subtracting from the speech spectrum the loss introduced by the A-to-B path, established as STC 20, and that curve is super-

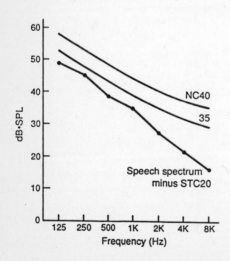

Figure 17-6. System analysis.

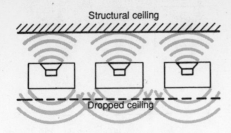

17-7. A plenum loudspeaker

Structural ceiling

Dropped ceiling

imposed on the NC 35 and 40 curves. Under these conditions, we have exceeded the zero dB signal-to-noise ratio that was the minimum goal of the system. It is quite likely that a masking noise corresponding to NC 35 would work. However, the designer might opt for NC 40 in order to allow for raised speech levels originating at A, since they would generally be 5 dB greater than the average levels.

We hasten to state that the specification and final adjustment of systems such as these require much expertise, and these are matters best left to qualified acoustical consultants who have had considerable experience in the area.

Electro-acoustical Considerations

If a loudspeaker array is intended solely for speech privacy, then it may be located in the plenum above the dropped ceiling, with the individually enclosed loudspeakers facing upward, as shown in Figure 17-7. This will ensure excellent diffusion of the masking noise spectrum at ear level in the office space, but the high-frequency equalization requirements might be severe.

More usually, we will see such systems incorporated with general paging requirements, and in these cases the loudspeakers will be mounted in the conventional downward facing manner. Loudspeaker density is important, inasmuch as the spectrum should not appear to vary as people walk around in the area covered by the system. The procedures outlined in Chapter 14 regarding coverage requirements should be applied here. Wide-angle transducers should be used, and this implies coaxial designs if 200 mm (8 in) diameter devices are specified. The –6-dB overlap angle requirements should be extended to 8 kHz, and the analysis should be made at ear level for the normal standing adult.

Absolute electrical stability of the system is essential. Nothing is more likely to be a source of complaints than a poorly designed or operated speech privacy system. Some of the considerations are:

1. Adequate powering. Shaped random noise spectra have fairly high crest factors (8-to-10 dB), and adequate amplifier power must be specified. Amplifiers should be operated well within specified limits.

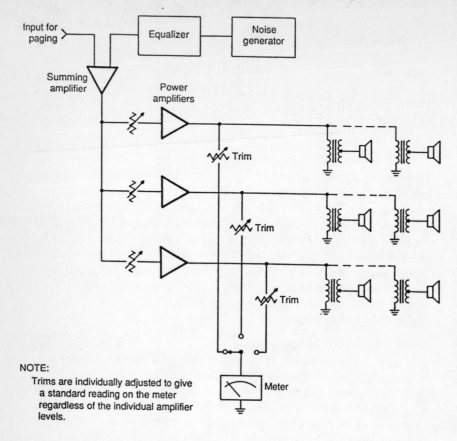

Figure 17-8. Electrical flow.

2. Spectrum shaping. Only the most stable noise sources should be used, and reliable one-third octave equalizers should be used for shaping the masking spectrum. Metering of amplifier output levels should be routinely performed.

3. Dual channel noise sources. It has been observed that feeding alternate loudspeakers with different noise generators, each equalized the same way, enhances diffusion of the masking noise.

4. System operation. The system should be designed for continuous operation and left on most of the time. It should be thought of as, say, the air conditioning or heating, turned on well before any workers arrive and off after all workers have left. It should never be interrupted during the work day. The paging function should not result in any diminution of the noise masking function.

An electrical flow diagram is shown in Figure 17-8.

Recommended Reading
1. Altec Technical Note Number 227A.
2. O.L. Angevine, "Comment on 'Speech Privacy,'" *db Magazine*, July/August 1985, p. 4.
3. W.J. Cavanaugh, W.R. Farrel, P.W. Hirtle, and B.G Watters, "Speech Privacy in Buildings," J. Acoustical Society of America, Vol. 34, p. 457 (1962).
4. R. Farrell, "Masking Noise Systems in Open and Closed Spaces," J. Audio Engineering Society, Vol. 19, Number 3 (1971).
5. J.A. Johnson, "A Simplified Articulation Index Calculation for Open-Plan Spaces," *Sound and Vibration*, Vol. 14, No. 6 (June 1980).
6. "Guide for Acoustical Performance Specification of an Integrated Ceiling and Background System," Geiger & Hamme, Inc., Ann Arbor, MI, GS-00-B-833 (PCCS).

CHAPTER 18

High-Level Sound Reproduction Systems

INTRODUCTION

In this chapter we will discuss the requirements for high-level sound reproduction in various environments. While specific examples will be given, most of this chapter will deal with the principles of analysis laid down in earlier chapters, which enable the designer to specify a system that meets detailed performance requirements.

GENERAL TECHNICAL REQUIREMENTS

Sound Pressure Levels and Headroom

The designer must determine the maximum sound pressure levels which a system is expected to deliver. A knowledge of system sensitivity ratings, maximum power input ratings, and available power will determine this. For most applications, existing loudspeaker systems can be specified. If the designer of the system intends to specify the system from raw components, then the principles discussed in Chapter 7 must be followed.

In multi-channel systems, the acoustical summation of levels should be based on the following equation:

$$\text{Level of ensemble} = L_1 + 10 \log(N) \tag{18-1}$$

In this equation, L_1 is the maximum level in SPL which can be produced a single loudspeaker at a given distance, and N is the number of loudspeakers. The assumption is further made that the loudspeakers are capable of producing the same output and that at times of maximum demand they are all contributing equally at the point of observation.

The question of absolute level requirements should be answered early in the analysis. Some confusion may exist between average and peak power ratings of loudspeakers. Most professional loudspeakers can withstand momentary inputs some ten times the rated power input provided that the duty cycle is short enough. The assumption is further made that the input signal is not so low in frequency as to require that the voice coil execute excursions beyond its mechanical limits. If a manufacturer gives a sine wave power rating for the loudspeaker, the duty cycle can be assumed to be a continuous one over the stated frequency range.

The same loudspeaker may have a program rating some three-to-five dB greater. Here, the conservative assumption is made that program peaks will be only three-to-five dB greater than average program input and that the loudspeaker voice coil will be able to dissipate sufficient heat between peaks so that there is no adverse temperature rise.

Biamplification, or even triamplification, is desirable in any high level music reproduction system for reasons explained in Chapter 7.

Power Bandwidth and Power Response Considerations

Many permanent high-level sound reproducing systems are installed in large spaces which have fairly low reverberation times, and a quick acoustical analysis will usually indicate that most of the audience is well within the direct field. Under these conditions, it is best to establish maximum output levels based entirely on direct field calculations. Any contribution by the reverberant field will then be a bonus.

Considerations of power bandwidth are important, and the designer must know over what bandwidth the loudspeaker will handle a given input power. A knowledge of the intended program spectrum is helpful. Most classical and jazz musical spectra roll off at high frequencies (see Chapter 7), and this will reduce the HF power bandwidth demands accordingly. However, much modern rock music has essentially a flat power spectrum, and a system analysis must be made assuming that acoustical output requirements at 10 and 12 kHz will be as great as in the mid-range.

The same considerations hold for power response. Today, it is comparatively easy to design systems which have smooth power response, since modern components exhibit uniform coverage over a wide frequency range. Review the discussion presented in Chapter 11.

Some degree of broadband equalization may be desirable in fine tuning a reproduction system. Certain aspects of power response may be addressed through equalization in the frequency range up to 300 or 400 Hz. More often, equalization will be helpful in precisely matching the inter-channel response of multi-channel stereophonic systems, compensating for acoustical asymmetry in the space.

Coverage Requirements

Slight toe-in of flanking loudspeakers in a stereophonic array will improve stereophonic localization over a large portion of the audience area. The data of Figure 18-1 illustrates this.

Front to back coverage in large rooms can be improved by aiming uniform coverage horns over the heads of the front listeners. In so doing, a trade-off can be made between inverse square losses and off-axis horn pattern losses. Details are shown in Figure 18-2.

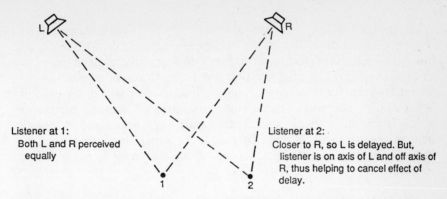

Figure 18-1. Toe-in of a loudspeaker for better stereophonic coverage.

Acoustical Considerations

In some large spaces, echoes, or discrete reflections, are often a problem, and they should be eliminated through proper surface treatment. In motion picture theaters floor-to-ceiling double-fold velour drapes are quite often used on side and back walls to render the space quite dead and free of reflections. A sense of ambience in the motion picture theater is generally created through the surround loudspeaker system rather than through natural reverberation in the theater.

The motion picture screen itself is quite reflective at high frequencies, so any reflection off the back wall is apt to result in a front-to-back flutter echo in the room.

Problems in recording studio control rooms are usually more complex in nature and are best left to those acousticians who specialize in them.

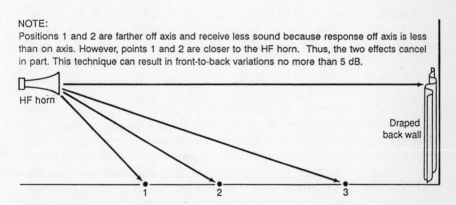

Figure 18-2. Aiming uniform coverage HF horns over listeners heads provides a trade-off between inverse-square and off-axis attenuation.

The designer can avoid many acoustical problems by suitably mounting the loudspeakers. Ideally, the loudspeakers should be flush mounted into the walls, as opposed to being free standing or hanging on cables or chains. The flush mounted position gives smooth response down to the lowest frequencies, whereas a hanging loudspeaker will likely exhibit peaks and dips at wavelengths equal to or greater than the perimeter of the front baffle. Such considerations are very important in motion picture theaters and in recording control rooms. They are of far less importance in a discotheque, where critical listening is not a prime concern.

A MOTION PICTURE SYSTEM

Basic Requirements

Since about 1980, motion picture reproduction technology has favored simple vented LF systems and uniform coverage HF horn systems. This is in distinction to the vented LF horn and HF multicellular horn systems which were the mainstay of motion picture sound reproduction for more than forty years.

Figure 18-3 shows a five-channel system for a motion picture theater. The physical characteristics of the space are:

$$
\begin{aligned}
\text{Volume} &= 5660 \text{ cubic meters} \\
\text{Surface area} &= 1100 \text{ square meters} \\
\text{Reverberation time (mid-band)} &= 0.5 \text{ seconds} \\
\text{Width} &= 22 \text{ meters} \\
\text{Height} &= 7.3 \text{ meters} \\
\text{Depth} &= 35 \text{ meters}
\end{aligned}
$$

The system consists of five channels behind the screen, and the loudspeaker ensemble is expected to produce peak levels at the rear of the house in the range of 105-to-110 dB-SPL. Sub-woofers are expected to produce sound pressure levels of 112-to-115 dB in the 20-to-30 Hz range, and an array of surround loudspeakers is expected to produce mid-band levels of 105 dB in the house.

In order that patrons at the front of the seating area perceive good stereophonic reproduction, loudspeakers having a 90-degree horizontal coverage angle will be required. Many companies provide systems specifically designed for motion picture application, and it is sensible to explore the catalogs for systems that directly meet these requirements. The systems shown in Figure 18-3 make use of HF uniform coverage horns with nominal coverage angles of 90 degrees horizontally and 40 degrees vertically. The LF sections of these systems have been augmented by doubling the array from two transducers to four.

Figure 18-3. A modern motion-picture loudspeaker system (JBL photo).

Acoustical Calculations

We will now make an estimate of the direct-to-reverberant ratio at the back of the house. The HF horn has a DI of 11 dB in the 2 kHz range, and a sensitivity of 113 dB-SPL, one watt at one meter. With input power of 1 watt, the direct field level at a distance of 35 meters will be:

Level = 113 – 20 log(35) = 82 dB-SPL

Using equations 13-2 and 2-23, we obtain the values of $\bar{\alpha}$ and R of 0.8 and 4400 square meters, respectively, for the space. The acoustical power into the space is calculated by Eq. 4-5 to be 20% of the electrical power input of one watt, or 0.2 watt.

Using Equation 2-25 we can then calculate the reverberant level to be 83 dB-SPL. Thus, the rear of the house is just about at the critical distance from one of the loudspeakers.

Critical distance may be determined from Equations 2-11 and 2-24 as 41.6 meters, and this is consistent with the previous observation.

The LF part of the system consists of four 380 mm (15 in) LF transducers arrayed vertically. The sensitivity of the array is 103 dB-SPL, one watt at one meter, and available power is 100 watts per transducer.

Peak outputs of 98 dB-SPL can be reached at the rear of the house with full power to the LF array.

The HF part of the system can develop direct field levels of 98 dB at the rear of the house when powered with 40 watts. When both HF and LF systems are driven under these conditions, the resulting level would be three dB greater or 101 dB-SPL, since the crossover frequency of 500 Hz is roughly at the geometric mean of the frequency range. (See Chapter 6 and note Equation 6-1.)

Using Equation 18-1, we observe that the ensemble of five channels will be about 7 dB greater, giving a total of 108 dB-SPL at the rear of the house. These are of course absolute peak levels and would be rarely, if ever, reached. More likely, we would not see levels exceeding 100 dB-SPL in the rear of the house.

In a space as acoustically absorptive as this, there is no diffuse reverberant field. Rather, the attenuation of sound with distance is likely to follow the quasi-steady-state sound field equations given in Chapter 1. In this event, the reverberant level will be even less than our present calculations give, and this is all the more reason to rely on direct field calculations altogether.

Coverage Requirements

As can be seen in Figure 18-3, the HF horns are aimed straight back. They are located at a height of 5.5 meters above the floor, and, demonstrating the effect shown in Figure 18-2, the front-to-back level variation is relatively small. The display shown in Figure 18-4 was generated by the CADP central array design program developed by JBL. It shows the normalized direct field coverage provided by the center HF element at 2 kHz. Note that the variation is ±2 dB over most of the house. If the HF horns are aimed downward, then there will be a "hot-spot" in the center of the house, with rapid fall-off beyond that point.

Equalization Requirements

It is customary for the screen loudspeakers to be equalized to match the response curve given in ISO Bulletin 2969. That response is shown as the dashed curve in Figure 18-5. Superimposed on that curve is the actual response of the center channel adjusted for flat power response and measured at a distance about two-thirds back in the house. Note that over the bulk of the range the unequalized curve departs from the ideal "house" curve by no more than 4 dB. Equalization is of course desirable to ensure that all five channels are virtually identical, and it is usually easy to match all screen systems to within ±1.5 dB of the ISO curve.

The HF losses observed in the dashed curve result from on-axis

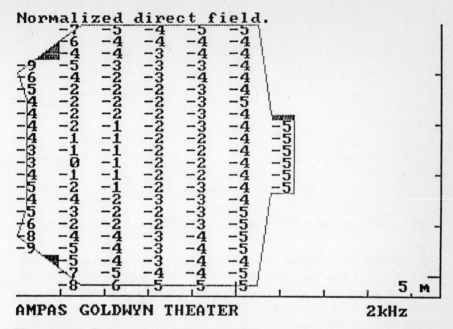

Figure 18-4. Simulated coverage in a motion-picture house with a center-channel HF horn aimed straight ahead; loudspeaker located at center left (JBL Data).

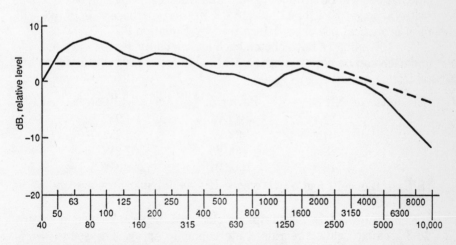

Figure 18-5. Theater equalization; the solid curve, center channel adjusted for flat power response, measured two-thirds into the house; the dashed line is ISO Bulletin 2969 and is recommended for theater equalization.

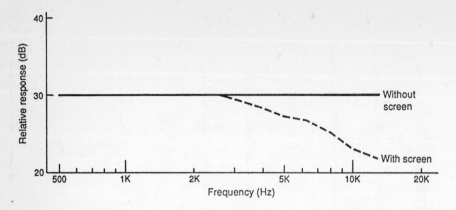

Figure 18-6. On-axis screen losses (8.5% opening; 0.5 mm vinyl material).

through-the-screen losses as well as atmospheric attenuation. Figure 18-6 shows typical on-axis screen losses for 8.5% opening in vinyl screen material which is 0.5 mm (0.020 in) thick. For off-axis positions, screen losses become fairly complicated and are beyond the scope of this analysis.

The Sub-woofer System

In calculating the requirements for sub-woofers, we use the method discussed in chapter 5. The assumption is made that reverberant field calculations will be sufficient and that mutual coupling will be effective.

For a target SPL of 112-to-115 dB in the reverberant field, we will need acoustical power as determined by Equation 2-25.

The LF units we have chosen each have a nominal efficiency of 2.1%, and each can be powered by 200 watts. We now construct the following table:

LF units	Eff (%)	Power in	Acoustical power	SPL*
1	2.1	200W	4.2W	96
2	4.2	400W	16.8W	102
4	8.4	800W	67.2W	108
8	16.8	1600W	268.8W	114

*SPL = 126 + 10 log(W/R), where R = 4400 square meters and W is acoustical power.

Eight sub-woofers were implemented, and they can be seen in Figure 18-3 located below the raised stereophonic array. The sound levels measured in the house were quite close to the targeted values, taking into account the characteristic peaks and dips of standing waves in the room.

The Surround System

The surround system consists of 12 loudspeakers located to the rear of the side walls, the back wall, and the rear of the ceiling. Each system is powered by 60 watts, and each has a sensitivity of 97 dB-SPL, one watt at one meter.

We will for convenience assume that a typical listener is at a mean distance from each surround loudspeaker of 10 meters. Under these assumptions, a level of 95 dB-SPL will be developed at the average listener. Using Equation 18-1, the ensemble of 12 loudspeakers will produce a level at the average listener of 106 dB-SPL.

Electrical Considerations

Figure 18-7 shows an abbreviated electrical layout for the motion picture system we have just described.

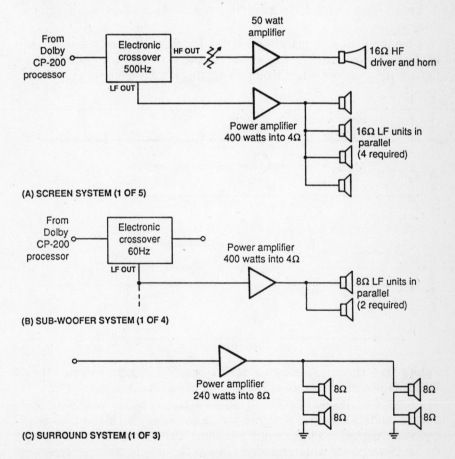

Figure 18-7. A theater system electrical diagram.

A RECORDING CONTROL ROOM MONITORING SYSTEM

General Considerations

It is recommended that the neophyte not attempt to design a control room monitoring system. There are many pitfalls, and there are so many excellent monitor loudspeakers to choose from. The control room itself is no simple thing as regards layout and acoustical treatment, and that complex subject is not one which this handbook addresses.

Among the fine points in monitor design which will have already been addressed by competent systems designers are:

1. Specific directional characteristics of transducers and their cross-over frequencies.

2. Detailed response shaping in the dividing network.

3. Physical and electrical time and phase alignment between transducers. (See Figure 7-16.)

4. Baffle layout and various boundary conditions affecting the transducers.

Some smaller systems are designed for flat LF response with the loudspeaker mounted in free-space (4π), and others are designed for flattest LF response when mounted against a solid boundary (2π) con-

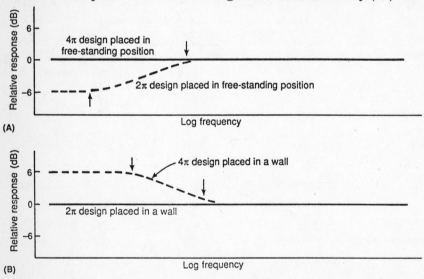

Figure 18-8. Basic boundary conditions for monitor loudspeakers. **(A)** The surface areas of a sphere is $4\pi r^2$; therefore a 4π boundary is one which requires the loudspeaker to radiate into all of space, or in all directions; a loudspeaker thus designed will be essentially flat at low frequencies when placed in a free-standing position. **(B)** A loudspeaker designed for flat LF response when mounted in a wall is termed a 2π (half-space) design.

294

dition. Figure 18-8 details the nature of these mounting conditions. The transition points indicated by the arrows in Figure 18-8 are dependent on the loudspeaker baffle perimeter. At frequencies with wavelengths longer than the baffle perimeter, the loudspeaker is effectively radiating into 4π space, while at frequencies twice that and above, the loudspeaker will radiate into 2π space. Conventional LF alignments using Thiele-Small parameters assume 2π radiation.

System Performance Requirements

Let us assume that the recording engineer will be seated two meters from a stereophonic loudspeaker pair and that sustained maximum levels of 115 dB-SPL may be required of the loudspeaker pair. Electrical headroom should allow for another 6 dB over this maximum level, and power bandwidth should be flat from 40 Hz to 12 kHz.

First, we take a look at the specifications of typical LF transducers intended for monitor use. These will normally be in the sensitivity range of 93 to 94 dB-SPL, one watt at one meter, and typical continuous sine wave power ratings are in the 150 watt range. Our first calculation tells us that 150 watts into such a transducer will produce a direct field level of about 109 dB-SPL at two meters.

A typical HF horn-driver combination will have a one-watt, one-meter sensitivity of 109 dB-SPL, and under conditions of matching the output of the LF section, the HF driver will require, in the mid-band, only 4 watts.

The acoustical load will be shared just about equally with the HF part of the system, and this will contribute another three dB, making a total of 112 dB. Further, a stereophonic pair will produce maximum levels three dB greater than a single loudspeaker, making a total of 115 dB-SPL at two meters. Thus, a single LF transducer per channel would seem to meet the specification.

Of course continued operation at maximum levels would be very uncomfortable for listeners, and it would certainly result in some degree of dynamic compression of the output, due to heating of the voice coil. (See chapter 5 for further discussion of this.)

Implementation of the system is shown in Figure 18-9. The system should make use of a horn-loaded HF section in order to meet the output demands. The choice of a two-way system crossing over in the 1 kHz range or a three-way system with an added VHF transducer crossing over at 8 kHz is a matter of judgement on the part of the designer. Even with the choice of a uniform coverage HF horn, the two-way implementation would satisfy the design requirements, as we have shown in Figure 18-10. Here, we have shown the increase in drive power above the 4-watts nominal maximum, needed to correct for the natural power response fall-off of the HF driver.

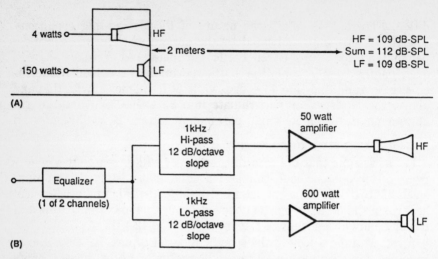

(A)

(B)

Figure 18-9. A studio monitor design. **(A)** Physical layout and acoustical summation. **(B)** Electrical implementation.

Manufacturer's recommendations for volume and tuning of the ported LF enclosure should be followed in order to ensure that the LF transducer can deliver its full power bandwidth at low frequencies.

Equalization Requirements

Critical studio monitoring systems are expected to conform to a specified response curve at the engineer's listening position, and matching between channels is critical. Figure 18-11 shows two equalization curves for monitors. Curve B is generally preferred, and a suggested tolerance is ±1.5 dB above 200 Hz. Below 200 Hz, the response may vary

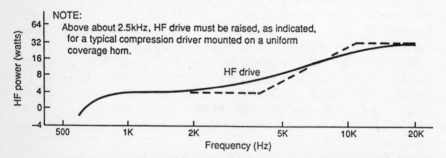

Figure 18-10. High-frequency power requirements for flat power response on a uniform coverage horn; the reference is full rated power output of the system.

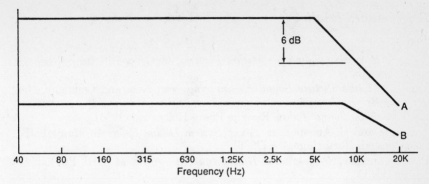

Figure 18-11. Monitor equalization contours.

by a somewhat larger amount. In any event, the two channels should be as well matched as is possible.

The measurement of system equalization is normally done with a Real Time Analyzer (RTA). Some engineers pick a single position just at the mixing engineer's ears to make the measurement, while others will take a space average in the general listening region. With careful attention to power response and room absorption, the amount of equalization required to adjust and match the systems will be minimal.

SOME CONSIDERATIONS FOR DISCOTHEQUE SYSTEMS

There is no limit to the imagination which has been shown in the special area of discotheque sound system design. These systems must be skillfully integrated into relatively large spaces, and the volume levels they must play at are truly awesome. Response down to the 20-25 Hz region is desirable, and HF power bandwidth can only be met through the implementation of VHF devices in multiples. It is rare for such a system to be less than four-way in its implementation, and each section should be independently powered.

Since there are always multiple loudspeakers per channel dispersed throughout the space, there is no consideration of stereophonic imaging *per se*. Often, the goal of the operators is to have the music literally split apart in frequency division, so that different parts of the spectrum appear to come from different directions. Further, the system is often integrated with the lighting so that certain lighting effects are synchronized with, or triggered by, the spectral nature of the music.

There are electroacoustical consultants with established track records in this special area, and they should be consulted early in the planning stages for discotheque systems.

Recommended Reading

Books:

Various, *Recording Sound for Motion Pictures*, McGraw-Hill Book Company, New York (1931).

Various, *Motion Picture Sound Engineering*, Van Nostrand Company, New York (1938).

L. Blake, *Film Sound Today*, Reveille Press, Hollywood (1984).

J. Eargle and G. Augspurger, *Sound System Design Reference Manual*, JBL Incorporated, Northridge, CA (1986).

D. Emenheiser, *Professional Discotheque Management*, CBI Publishing, Boston (1981).

M.Z. Wysotsky, *Wide-screen Cinema and Stereophonic Sound*, Hastings House Publishers, New York.

Articles:

J. Eargle, "Equalizing the Monitoring Environment," J. Audio Eng. Soc., Vol. 21, No. 2 (1973).

J. Eargle and R. Means, "A Microcomputer Program for Determining Loudspeaker Coverage in Motion-Picture Theaters," J. Soc. Motion Picture and Television Engineers, Vol. 93, No. 8 (1984).

J. Eargle, J. Bonner, and D. Ross, "The Academy's New State of the Art Loudspeaker System," Journal, SMPTE, Vol. 94, No. 6 (1985).

M. Engebretson and J. Eargle, "Cinema Sound Reproduction Systems," Journal, SMPTE, Vol. 91, No. 11 (1982).

CHAPTER 19

Overview of Sound Reinforcement in the Theater

INTRODUCTION

Until recent years, there was little sound reinforcement in either the legitimate theater or the musical theater. Traditionally, most of the performance venues have been small, and actors or singers were expected to be able to project their voices naturally.

In recent years, as road shows have proliferated and as performances have played in larger venues, the need for some degree of sound reinforcement has become an important one.

Pick-up techniques can vary from relatively subtle footlight microphones all the way to wireless microphones worn by the actors. In the musical theater, it has become customary to amplify the pit orchestra, although this is hardly necessary, and many a patron has complained of the often ear-splitting sound in the theater.

From the beginning of electroacoustics, it has been quite appropriate in the theater to create some special effects electronically. Some examples here are off-stage effects or voices, such as the Witch of Endor or the ghost of Hamlet's father.

In this chapter, we will discuss the specific pickup techniques used in the theater and outline the requirements for large scale reinforcement and handling of special effects.

BASIC REQUIREMENTS IN LARGE HOUSES

Requirement for Acoustical Gain

In general sound reinforcement, there may be a need for considerable acoustical gain, and system layout often stresses this need. In the legitimate theater, there may not be a need for significant gain, inasuch as the surroundings are relatively quiet and the actor's voices usually project well. What is needed more often than not is just a little more clarity of articulation, and this can be arrived at through subtle emphasis of high frequencies in a relatively low gain system.

Figure 19-1A shows the basic layout for a simple theater reinforcement system which requires little supervision other than basic setting of operating levels. The system is composed of three channels and as such

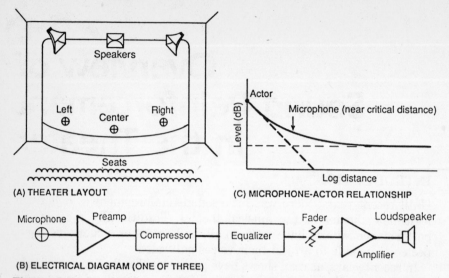

(A) THEATER LAYOUT

(C) MICROPHONE-ACTOR RELATIONSHIP

(B) ELECTRICAL DIAGRAM (ONE OF THREE)

Figure 19-1. A simple system for theater sound reinforcement. **(A)** Theater layout. **(B)** An electrical diagram of one of the three. **(C)** Microphone-actor relationship.

probably will produce a very natural effect. The electrical diagram is shown at B, and the typical actor-microphone relationship is shown at C. Boundary type microphones are the best to use for this application.

If operation of such a system is attempted at too high levels, there will be a tendency for reverberant sound to be amplified—and this may defeat the original purpose of the system. In general, no more than 6 dB of gain should be expected of the system, and this is usually more than enough. Care must be taken that extraneous noises, such as footfalls, are not unduly emphasized. Rolling off low frequencies is common in the operation of such systems as this.

In the case of musical productions in venues seating 1500 or more patrons, the gain requirement may be considerable, and wireless microphones worn by the actors will provide all the gain necessary because of the quite small D_s distance. The drawback of these microphones of course is that they also pick up the rustling of clothes and other extraneous noises. They are not always free of electrical interference effects.

Stereo versus Mono

Single-channel central array systems are the most common fixed installations in theaters. For speech reinforcement, they are preferable to two-channel side arrays, which are often brought into a theater for a particular road show. Most effective of all are the multi-channel overhead arrays, which are rarely installed on a temporary basis, and which are permanent features of too few houses.

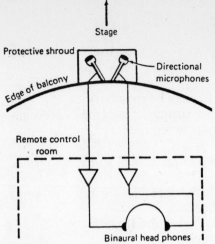

Figure 19-2. Remote monitoring of a reinforcement system using headphones.

Monitoring the System

While the system shown in Figure 19-1 needs little or no monitoring once levels have been set, typical theater systems are more complex. The ideal position for the operator's console is mid-house, and the front center of the balcony is preferred because of excellent sight lines.

The operator monitors the actual acoustical levels in the house, and it is his responsibility to maintain an appropriate level for the program as well as sufficient feedback stability margin.

An alternative monitoring method in fixed installations is shown in Figure 19-2. Paul Veneklasen has favored this approach in many of his theater system designs.

Stage Monitoring Requirements

Stage monitoring, or foldback, provides some degree of reinforced sound purely for the benefit of the performers. While few seasoned actors will require foldback, it is commonly provided for choruses and some vocalists. The benefit is a psychological one, and it often enables singers to maintain accurate pitch more easily. Figure 19-3 shows details of this.

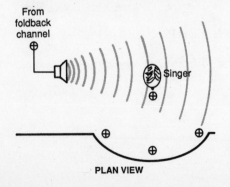

Figure 19-3. Foldback of reinforcement for singers.

301

SYSTEM IMPLEMENTATION

A large comprehensive theater system will address the following requirements:

1. Inputs:

   Low level: microphones (up to 32 may be required)

   High level: from tape recorder or turntable; used for background music or recorded effects (up to 8 may be required)

2. Output busses assignable to:

   Recording channels (depending on application, 4 or more may be required)

   Reinforcement channels (up to 6 may be required; 5 in central array and one delayed under-balcony channel)

   Foldback channels (up to 4 may be required)

3. Ambience and special effects busses

   Up to 8 required, each capable of being fed a unique mix of delayed and reverberated sound

4. Input signal processing:

   Equalization

   Limiting/compression

   Delay/reverberation send functions

5. Output signal processing:

   Equalization

   Limiting/compression

   Panoramic potentiometers (panpots) for manual steering of signals

   Delay/reverberation return functions

Above all, a comprehensive theater system will provide sufficient flexibility for reconfiguration as required by ever changing program demands.

Bibliography:

1. H. Burris-Meyer, et al., *Sound in the Theatre*, Theatre Arts Book, New York (1979).

2. J. Eargle, *The Microphone Handbook*, Elar Publications, Plainview, NY (1982).

3. Various, *Sound Reinforcement*, (compiled from the pages of the Journal of the Audio Engineering Society, New York, 1978).

CHAPTER 20

Live Music Reinforcement: An Overview

INTRODUCTION

Live music reinforcement spans the gamut from the simple demands of a club or cabaret environment up to large-scale reinforcement of rock groups in large arenas or outdoor spaces. Since the program is music, matters of power bandwidth and broad coverage take precedence over the concerns of speech intelligibility.

A high-level music reinforcement system differs from a speech reinforcement system in the following ways:

1. Very short D_s (source to microphone) distances. Many instruments are picked up by contact microphones or by direct electrical connection, and in those cases, D_s is virtually zero. Feedback is rarely a problem if proper operating procedures are observed.

2. High acoustical power output and extended power bandwidth. Patrons demand and are willing to pay for high sound pressure levels, often in the 110 to 115 dB range. Ample loudspeakers and amplifiers must be available to produce these levels cleanly over the entire seating area.

3. Rich use of signal processing and effects. The expectation at a rock concert is that the group sound substantially like the phonograph records which may have preceded it. Reverberation and other signal processing may be needed to accomplish this.

4. Ruggedness. Portable systems must be assembled and disassembled many times during the course of a year. Custom designed enclosures and carrying cases are standard, and they are expensive.

5. Efficiency of assembly. Time is money, and music reinforcement system designers have adopted heavy duty hardware for both mechanical and electrical connections. Preassembled cables allow speedy set-up and break-down of jobs.

6. Modular concept in acoustical layout. Most music reinforcement companies have designed their own full-range acoustical modules. These can be specified in quantity for given coverage and level requirements.

7. Requirements for stage monitoring. In reality, there are two systems: one to cover the house, and another, called *stage monitoring* or

foldback, to enable the musicians to hear themselves and each other. This is important for the sake of good ensemble playing.

8. Special measures for large venues. Delayed arrays are often required in large venues, and HF equalization may be required to compensate for HF air losses. Multichannel reinforcement is common.

9. Rigging requirements. Specialists in the area of music reinforcement must be very aware of all liability assumed in system rigging. The detailed requirements for safety implicit in this are extremely important and beyond the scope of this handbook.

SMALL SYSTEMS FOR CLUB USE

Many manufacturers make systems intended for portable, small-scale music reinforcement. Typical examples are shown in Figure 20-1A and B.

Such systems are generally characterized by:

1. Rugged construction. Plywood is preferred over particle board; finishes should be durable and scuff resistant. Some manufacturers use industrial grade carpet on the outside of the enclosure, since it can take much abuse and still maintain an acceptable appearance. Handles are usually incorporated into the sides of the enclosures, and casters are often included with larger systems. Covers and carrying cases may be optional.

2. Acoustical characteristics. The transducers used in these systems are often those intended for musical instrument (MI) applications. The characteristics here are relatively high efficiency (5 to 8%) and the tailoring of given systems to specific musical applications. For example, there may be systems intended for lead guitar, keyboard, vocal, or bass guitar use, and so forth.

3. Electrical interface. The usual input connector is the simple 2-conductor phone plug. It is used as shown in Figure 20-1C, and paralleling of jacks enables multiple loudspeakers to be conveniently connected. The user must be constantly aware of the reduced impedance load which multiple parallel loudspeakers will cause.

Notes on Application

With proper planning, components such as those shown in Figure 20-1 may be mounted permanently in clubs of moderate size. Care must be taken not to exceed the power input limitations of such systems. Line arrays (column loudspeakers) may be best for vocal applications, and they should be placed as far to the sides as is convenient to minimize feedback. (See the discussion on column loudspeakers in Chapter 21.) Above all, manufacturer's specification sheets should be carefully studied and followed.

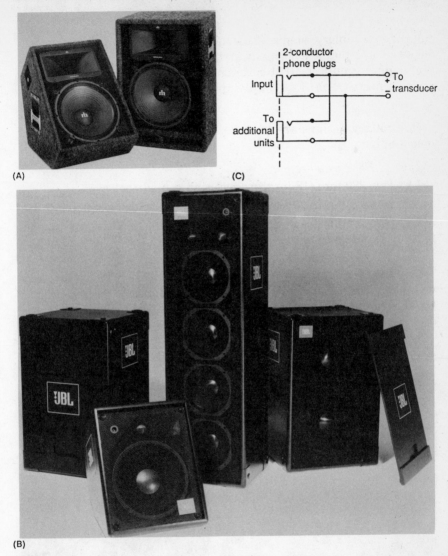

In the diagram (C):

2-conductor phone plugs

Input

To additional units

+ To transducer

(A)

(C)

(B)

Figure 20-1. Systems for club use. **(A)** Renkus-Heinz photo. **(B)** JBL photo.

COMPONENTS FOR LARGE-SCALE MUSIC REINFORCEMENT

The Modular Approach

While it is typical of designers of fixed installations to specify elements covering specific parts on the spectrum and orient them individually, the music reinforcement specialist tends to use integral pre-designed

305

full-range modular systems. These modules are specified in the quantity required, and they are normally arrayed overhead, or to the sides of the performing group, for best coverage of the seating area. Figure 20-2 shows a photograph of a typical application.

Module Design

Proprietary designs are the rule, and four examples are described generically:

System 1. Four-way design:

LF: Two 380 mm (18 in) transducers in vented enclosure.

MF: Four 250 mm (10 in) transducers in sealed enclosure.

HF: Two compression drivers with 100 mm diaphragms coupled to short exponential horns.

UHF: Two ring radiators.

Crossover frequencies have been selected to distribute power bandwidth requirements evenly throughout the frequency range. At a one meter reference distance, this system in multi-amplified mode can produce program levels in excess of 130 dB-SPL.

Figure 20-2. Music reinforcement at the Hollywood Bowl (JBL photo).

System 2: Four-way design:

LF: Four 460 mm (18 in) transducers in vented enclosure.

MF: Two 380 mm (15 in) transducers in sealed enclosure.

HF: One compression driver with 100 mm diaphragm coupled to short exponential horn.

UHF: Four ring radiators.

Crossover frequencies have been selected to distribute power bandwidth evenly over the frequency range. In multi-amplified mode, the system can produce program levels at a reference distance of one meter in excess of 129 dB-SPL.

System 3: Four-way design:

LF: Three 460 mm (18 in) transducers in vented enclosure.

MF: Two 300 mm (12 in) transducers in horn loaded configuration.

HF: One HF compression driver with 100 mm diaphragm mounted on short 90-degree by 40-degree horn.

UHF: Two ring radiators.

Crossover frequencies have selected to distribute power bandwidth evenly over the frequency range. In multi-amplified mode, the system can produce program levels at a reference distance of one meter in excess of 129 dB-SPL.

System 4: Four-way design:

LF: One 460 mm (18 in) transducer in horn loaded enclosure.

MF: Two 300 mm (12 in) transducers in horn loaded enclosure.

HF: One 100 mm diaphragm HF compression driver mounted on proprietary horn.

UHF: Two ring radiators.

Crossover frequencies have been selected to distribute power bandwidth evenly over the frequency spectrum. In a multi-amplified mode the system can produce program levels in excess of 130 dB-SPL at a reference distance of one meter.

While the above systems differ in detail, they are remarkably similar in their individual power output ratings. They are all fairly similar in size, and angular coverage at middle and high frequencies is fairly similar.

They are all specified according to similar rules, and overall coverage requirements are met through the stacking and splaying of loudspeaker modules.

Stage monitors are usually of a different design than the modules used in the house system, and their quantity and placement is determined by the stage layout.

SIGNAL FLOW

Figure 20-3 shows a typical signal flow diagram for a large rock reinforcement system. Note that the stage monitor feeds (whether from direct or from microphones) are routed through microphone splitters. One set of outputs goes to the monitor mix console, and the other set goes to the main reinforcement console.

The two mixes may be significantly different. Players on stage need to hear a mix which helps them to maintain good ensemble; and of course each player may want a monitor mix of his own. The house feed represents the sound of the group which the patrons expect, and if the performing group is known through its recordings, the house mix is expected to sound much the same.

Signal processing, including time delay, reverberation, noise gating, and compression, will be required in order to satisfy all program demands for both mixes. The house feeds may be mono, stereo, or multi-channel as desired or as dictated by the performance venue.

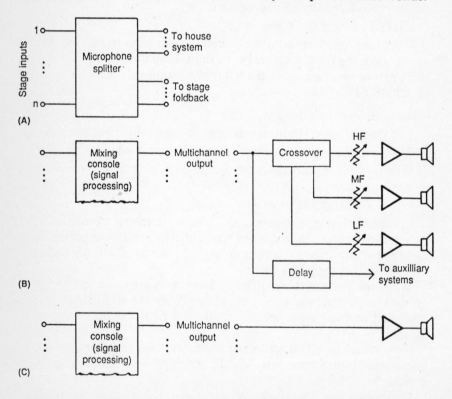

Figure 20-3. An electrical flow diagram. **(A)** Input. **(B)** House system. **(C)** Stage system.

Both house and monitor mixes will require their own operators, and both must operate from appropriate positions in the house.

For fool-proof and efficient set-up, it is necessary that positive interconnections be used between all electrical components. XLR connectors are generally used for all signal sources and for console output to both amplifiers and external signal processing gear. Amplifier outputs are usually routed to four- or six-pin connectors, depending on whether the system is bi- or tri-amplified.

DETERMING COVERAGE REQUIREMENTS

The rules which determine the number of modules to use for a given job are not hard and fast, and most companies involved in large-scale music reinforcement have worked out their own empirical formulas. A great deal of what is done is based simply upon what has worked in similar venues in the past. Economic factors are important; remember that a 3-dB increase in overall output level capability will require a doubling of acoustical and electrical resources.

LOUDSPEAKER ARRAYS

The system may be mono, stereo, or multi-channel, as required by the music or the venue. For very large outdoor work, secondary arrays may be used, suitably delayed, so that the effects of inverse square losses can be compensated. In this regard, the principles of time delay, as discussed in Chapter 14, are used. The system shown in Figure 20-2 shows a side array (far right in the photograph) delayed to provide adequate coverage in the 180 meter (600 ft) deep space.

ELECTRICAL POWER CONSIDERATIONS

When thousands of watts of audio power are required, along with even greater power demands for lighting, large concerts will always require that special electrical service be provided. Power is normally distributed for industrial use in a three-phase 240 volt circuit. Few audio components can handle 240 volts, and the potential must be stepped down to 120 volts. Further, most components are single-phase, and this will require that the three legs of the three-phase service be equally loaded so that the neutral point in the three-phase distribution will be maintained close to zero potential.

Electrical Noise and Radio Frequency Interference (RF)

There is often little time to solve noise problems once a music reinforcement system has been set up. It is therefore necessary that every precaution be taken to eliminate noise and RFI before they have a chance to develop. Balanced lines are essential at all console signal inputs,

309

and they should be used at line output levels as well. Microphone cable varies considerably in its immunity to noise pickup, and only the best cable should be used. The cable configuration known as star-quad is recommended for microphone use because of its high immunity to electromagnetically induced noise.

It is good engineering practice to run signal wiring as far as possible from power lines for lighting. The control systems used in lighting generate considerable interference which may show up as annoying buzzing in the audio signal if precautions are not taken.

Suggested Reading

The various trade magazines covering the fields of recording and sound reinforcement have over the years presented detailed descriptions of music reinforcement systems, both large and small. It is worth the time of anyone interested in the art to research the last five year's worth of publications in this area and digest them. As is often the case, the really good practitioners in the art are often too busy to document their own work.

In the United States, the following magazines are recommended: *db*, *Recording Engineer/Producer*, and *Mix*.

CHAPTER 21

Line Arrays and Sound Columns

INTRODUCTION

While complex line arrays have attained a relatively high status in Europe, they have not until recently captured the interest of American designers. The commercial sound column is the simplest line array, and the device is often pressed into service when it should not be. Under the right circumstances, however, it can represent a logical component choice.

In recent years, complex line arrays have been designed which handle large amounts of power and which provide excellent coverage for both speech and music reinforcement. In this chapter we will give an overview of these devices and their applications.

SIMPLE LINE ARRAYS

As we observed in Figure 4-25, a simple line array produces a polar pattern at short wavelenghs which exhibits considerable narrowing and lobing. The range over which the array may exhibit a fairly well controlled pattern is a complicated function of the number of elements in the array, the spacing between them, and the matching of their individual responses. In a typical line array consisting of six to eight transducers, the resulting frequency range over which smooth pattern control is maintained may be no more than three octaves.

Response Tapering

What is required to extend the HF coverage is reducing the effective length of the array at high frequencies so that off-axis lobing will be minimized. This is known as *response tapering*, and it involves rolling off HF response of the outer elements progressively with rising frequency. Figure 21-1 shows three ways in which this can be done. Electrical roll-off, shown at A, is most common. Acoustical roll-off, as shown at B, has been used, and the "barber pole" arrangement, shown at C, makes use of the increasing directivity of the individual components with rising frequency. In any given direction, the entire array is functioning at low frequencies, while at high frequencies only those three or four elements pointing in a given direction are effective in that direction. The barber pole array must be specified for applications requiring relatively wide horizontal coverage.

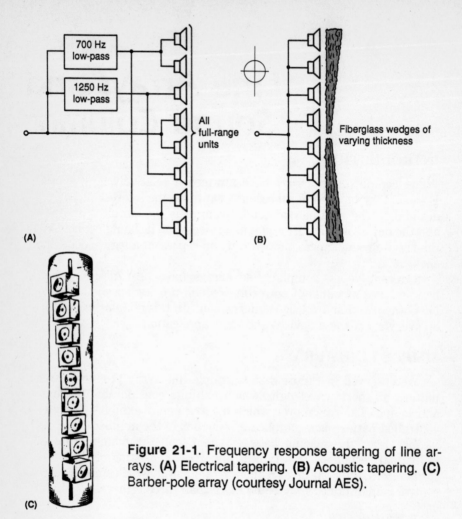

Figure 21-1. Frequency response tapering of line arrays. **(A)** Electrical tapering. **(B)** Acoustic tapering. **(C)** Barber-pole array (courtesy Journal AES).

Curved Arrays

If a line array is curved, as shown in Figure 21-2A, improved HF response can be attained with less frequency tapering. A refinement of this is shown at B in which the curvature has been attained through progressive time delay. The approach is expensive, but it has an advantage over that shown at A in that its horizontal off-axis response will be smoother.

With modern computer modeling, all factors regarding response tapering, delay, and spacing can be optimized for coverage over a wide frequency range.

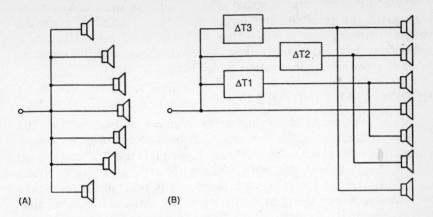

Figure 21-2. Curved line arrays (response tapering not shown). **(A)** Curved array. **(B)** Variable delay array.

NOTES ON THE USE OF COMMERCIAL SOUND COLUMNS

Sound columns can be quite useful if a reinforcement system designer keeps in mind their inherent limitations. We give below some rules to follow in their use and specification.

1. Power class. Keep in mind that sound columns are usually made up of no more than eight low cost small transducers. Power handling is limited, and they should be used only for speech reinforcement applications where demands below 125 Hz are minimal.

2. Grazing incidence is preferred. Sound columns may be used in spaces where the horizontal coverage distance is up to six or eight times the height of the column above the floor. In these cases, the sound column may be aimed straight back so that vertical off-axis characteristics tend to complement inverse square fall-off. (See discussion of this in Chapter 18.) We assume here that the off-axis response of the column has been maintained fairly smooth through tapering.

3. Adjuncts in a distributed, delayed system. While ceiling mounted transducers are required in distributed systems in rooms with relatively low ceilings, houses of worship will require wall loudspeakers. In such cases, the sound column may be the best choice, since the sound can be aimed where it is needed. Review the discussion of this in Chapter 14.

4. Gain considerations. If sound columns are used in fairly small spaces, place the sound column as far from the microphone as possible. If the podium is on the left, place the sound column on the right. While the effect is not entirely natural, it will allow more system gain before the onset of feedback. In addition, use as small a D_s (distance between the talker and microphone) as possible.

5. When possible, choose sound columns with as large and as many components as possible. The larger models will exhibit controlled vertical coverage over a wide range with minimal lobing. Horizontal response will be relatively wide at all frequencies, and this must be taken into account at the system design stages.

HORIZONTAL LINE ARRAYS

A special application of the line array is shown in Figure 21-3. The room is rectangular, and a line array has been placed over the proscenium from wall to wall. No electrical tapering is used. There are reflected images in both side walls, and the line array becomes, in effect, an infinite one. As a result, the listener at any point in the room perceives the source of sound as localized directly ahead of him, since that will always be the direction of the first arrival sound.

Such a system, designed by Klepper Marshall King Associates, has been installed in the concert hall at Peabody Conservatory in Baltimore, Maryland, and consists of more than fifty 200 mm (8 in) loudspeakers wired in series-parallel. The attenuation of direct sound, as illustrated in Figure 2-9, is 3 dB per doubling of distance. Such a system as this can only be specified for monophonic use, and it is intended for speech applications.

CUSTOM LINE ARRAYS

Today, acoustical designers have brought line array design to a fine art. Figure 21-4 shows details of a complex array. Note the relatively straight

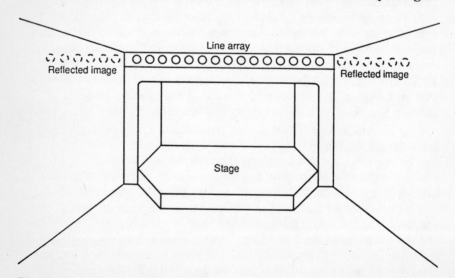

Figure 21-3. Horizontal line array.

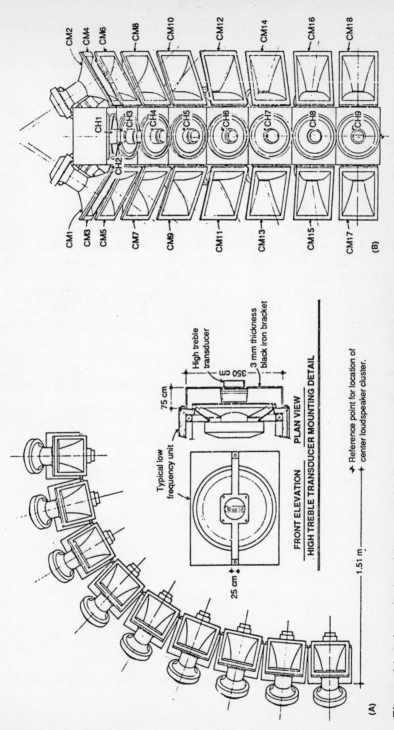

Figure 21-4. An articulated line array (Joong Ang. Daily News Building; Welton Becket Associates.) **(A)** The side view. **(B)** The front view.

lower portion of the array. This provides increased directivity toward the rear of the room. The front of the room is covered by the curved upper portion of the array. Elements aimed both left and right ensure broad coverage of the seating area.

It is customary in such arrays as this to power the components in several sub-groups so that drive levels can be adjusted without introducing loss into the amplifier output circuitry. This approach also allows fine tuning of the system *in situ*. Some questions regarding exact coverage can be conveniently answered after the system has been installed, avoiding costly re-arraying of the components.

FINAL COMMENTS

Some designers model line arrays using loudspeakers approximately one-fourth to one-eighth the size of the actual components. Such models are small and can be conveniently measured, taking into account wavelength scaling.

Line arrays can be complex and expensive, but if properly designed they have advantages of low distortion and precise coverage. In the future, they will undoubtedly become more popular in general sound reinforcement work.

Recommended Reading:
1. H. Kuttruff, *Room Acoustics*, pp. 268-271, Applied Science Publishers, London (1979).
2. G. Augspurger and J. Brawley, "An Improved Colinear Array," presented at the Audio Engineering Society Convention, New York, 8-12 October 1983; preprint number 2047.
3. W.J.W. Kitzen, "Multiple Loudspeaker Arrays Using Bessel Coefficients," Philips Technical Publications, Eindhoven, Netherlands.
4. D. Klepper and D. Steele, "Constant Directional Characteristics from a Line Source Array," *Journal Audio Eng. Soc*, Volume 11, Number 3, (1963).

INDEX

-A-

-E-

-F-

-G-

-M-

-N-

-O-

-P-

-Q-

-R-

-S-

-T-

-U-

-V-